IF YOU LOVE THIS PLANET

ALSO BY THE SAME AUTHOR

Missile Envy

Nuclear Madness: What You Can Do

IF YOU LOVE THIS PLANET

a plan to heal the earth

Helen Caldicott, M.D.

W. W. NORTON
& COMPANY
new york · london

I wish to thank Gabby Weiley
for her tireless typing of this at
times unwieldy manuscript and
Diane Civincione for its conception.

Printed in the United States of America
First Edition

The text of this book is composed in Bembo with the display set in Antique Olive Compact and Avant Garde Extra Light. Composition and manufacturing by the Haddon Craftsmen, Inc.
Book design by Charlotte Staub

Library of Congress Cataloging-in-Publication Data
Caldicott, Helen.
If you love this planet: a plan to heal the earth / Helen Caldicott.
p. cm.
Includes index.
1. Pollution. 2. International business enterprises—Environmental aspects. 3. Environmental protection. I. Title.
TD174.C33 1992
363.7—dc20 91-30322

ISBN 0-393-03045-8
ISBN 0-393-30835-9 (pbk)

W. W. Norton & Company, Inc.
500 Fifth Avenue, New York, N.Y. 10110

W. W. Norton & Company Ltd.
10 Coptic Street, London WC1A 1PU

1 2 3 4 5 6 7 8 9 0

I dedicate this book to all the birds,
animals, and plants of the planet.

Contents

IF YOU LOVE THIS PLANET

Introduction

> The only thing necessary for the triumph of evil is for good men to do nothing.
>
> —Edmund Burke

I first visited the United States in 1966, when I was a conservative young mother with three babies and a new medical degree. I took a job at Harvard Medical School, in the cystic fibrosis clinic of the Children's Hospital Medical Center, and worked part-time while continuing to care for my children. Because of the American films and television shows that I had grown up watching in Australia, I expected a gangster to be lurking behind every lamppost and the streets to be filled with neon lights and fast-food joints. Instead, I found the beautiful, orderly villages of New England and the quiet, reserved but deeply caring people of Down East and Boston.

But the years 1966–69 were years of political turbulence and violence. I witnessed the anti–Vietnam War movement, heard protest songs on the radio, watched flower children on television, and wept into my ironing as I listened to George Wald deliver an incredibly powerful and moving address on the day of the 1969 antiwar moratorium at the Massachusetts Institute of Technology.

In March of 1968, I heard the redoubtable Louis Lyons raging one night on the radio as he described the killing of Martin Luther King, Jr., exhorting all of us to march in the streets against such iniquity. A few months thereafter one sultry Boston summer morning, I turned on the television to see Bobby Kennedy lying bleeding from the head on the floor of a Los Angeles hotel. I found myself screaming at the TV, "Not again!" Several days later, I drove up to the Berkshires and listened in tears to the broadcast of his funeral from St. Patrick's Cathedral.

With a sense of dread, I saw Richard Nixon elected to the presidency that fall. Impelled by a growing desperation, I wrote to him about the cold war and the Antiballistic Missile Treaty and I am sure that he used part of my letter in his inaugural address. I even wrote to Senator Edward Kennedy about my deep concern about nuclear war and the importance of the upcoming ABM Treaty, and he actually replied with a thoughtful letter, which excited this rather naive young doctor from Down Under.

It was a thrilling time for me. Radicalized politically, I realized that democracy was a workable proposition, because the turmoil seemed to be igniting change. Anything, I thought, was possible.

In 1969, I returned to Australia and two years later successfully "took on" the French government, which was testing nuclear weapons in the atmosphere over some remote Pacific islands. The French tests violated the Partial Test Ban Treaty and created radioactive fallout in my small city of Adelaide, in South Australia. After nine months of a public education campaign warning about the medical dangers of strontium 90 and radioactive iodine in mothers' milk, 70 percent of my fellow Australians rose up and demanded that our government take action to stop the tests. The government did so, and the French were forced to test underground. I had been transformed into an activist in the United States, and my life was never to be the same again.

I have a deep regard for America as a land of excitement, change, and opportunity. It is also the country that will determine the fate of the earth. Why do I make such a sweeping statement? Because the United States is the wealthiest and militarily the most powerful country and because its powerful media penetrates into every corner of the world, establishing the models that most people wish to emulate. Thanks to its influence, millions of Chinese, Africans, Indians, and Latin Americans want cars, refrigerators, ice cubes in their Coke, air conditioners, and disposable packaging.

Because the U.S. population represents only 5 percent of the earth's people but uses 25 percent of the energy, this life-style is not an appropriate model for billions of other people.[1] Such extravagant living is the leading cause of ozone depletion, global warming, toxic pollution of the air, water, and soil, and nuclear proliferation. Each U.S. resident causes twenty to a hundred times more pollution than any Third World resident, and rich American babies are destined to cause a thousand times more pollution than their counterparts in Bangladesh or Pakistan.[2] Canadian citizens copy this lavish life-style of their southern neighbors, as do most citizens of the wealthy developed countries. Many Third World countries, too, and the now dismantled Soviet Union and China are on the verge of capitalism, and their people will demand this affluent life-style. But 5.4 billion people cannot possibly emulate the life-style of 250 million Americans and expect the planet to survive. Furthermore, the earth's population may increase to 14 billion within the next century.[3]

Some eminent scientists predict that if we do not act now to reverse the cumulative effects of global pollution, species extinction, overpopulation, and the ongoing nuclear threat, it will soon—possibly within ten years—be too late for the long-term survival of most of the planet's species, perhaps even *Homo sapiens*.

My vocation is medicine, and as a physician I examine the dying planet as I do a dying patient. The earth has a natural system of interacting homeostatic mechanisms similar to the human body's. If one system is diseased, like the ozone layer, then other systems develop abnormalities in function—the crops will die, the plankton will be damaged, and the eyes of all creatures on the planet will become diseased and vision impaired.

We must have the tenacity and courage to examine the various disease processes afflicting our planetary home. But an accurate and meticulous diagnosis is not enough. We never cure patients by announcing that they are suffering from meningococcal meningitis or cancer of the bladder. Unless we are prepared to look further for the cause, or etiology, of the disease process, the patient will not be cured. Once we have elucidated the etiology, we can prescribe appropriate treatments.

Unfortunately, the etiologies of the diverse diseases afflicting our planet are complex and difficult to face examined in fine detail. The initial, wondrous promises of capitalism and corporate free enterprise have not always led to a careful and responsible management of the earth's natural resources and treasures. And the ills of communism led in many cases to disastrous pollution and wanton neglect of nature.

In this book, I outline the diagnosis of planetary ills and then analyze the causes of these diseases. This involves a critical dissection of transnational corporations and their impact upon American society and the world at large. It also includes a brief analysis of the now defunct Communist system relating to pollution and ecological damage.

I am trained as a physician, and my approach to problems is thus always medical. I have therefore arranged this book using the criteria one would use to diagnose and adequately treat a patient. The order is as follows:

1. description of signs and symptoms (chapters 1–5)
2. diagnosis (chapter 6)
3. discussion of causes of the illness (etiology) (chapters 7–9)
4. prescription for a cure (chapter 10)

The first two years of medical school, devoted to the study of basic biological sciences, are often difficult, because a tremendous amount of information must be absorbed. But they are essential to the production of well-rounded clinicians who can practice the art and science of medicine. Similarly, the next five chapters of this book are packed with basic facts about the demise of our planet, but we must have these facts at our command in order to be adequately equipped to save the earth. As I have already said, I have a great regard for America and its people. I admire their resilience, creativity, and largess. What follows in this book may strike some readers as being harsh and overly critical. Yet I write out of love and concern—for the United States and for the planet.

Students at a college in Napa Valley recently asked me, "What do you think of American people." Having never encountered that question before in public, I searched for an honest and responsible reply. I had to say that Americans are the kindest and most caring people on earth. They desperately want to do the right thing, but they are just not sure what that is. Is it to fight a war in the Persian Gulf and support George Bush and feel patriotic? Or is it to care for the homeless and address the domestic issues of racism and economic inequality? Or is it to save the planet?

There are no easy answers to these questions, but I will attempt to find some in this book. Let us not be afraid to look at our own society in a critical fashion, because from a caring and rigorous analysis, we will fashion a cure for the dying earth.

If you love your country enough to cure its ills, you will be able to love and cure this planet.

1

Ozone Depletion and the Greenhouse Effect

There is a hole in the ozone layer. Ozone is being destroyed by chlorofluorocarbon (CFC) gases, which are like nuclear wastes—their effects are ongoing, and their lifespan in the ozone layer is about a hundred years. Since the hole was discovered, over ten years ago, ozone depletion has more than doubled.[1]

Before there was life on earth, there was no ozone. Ozone protects us from the lethal consequences of solar ultraviolet light, which kills cells. But as evolution began, single-celled plants relatively resistant to ultraviolet light started to grow and by the process of photosynthesis, which uses chlorophyl as a catalyst, absorbed carbon dioxide (CO_2) from the air and created oxygen (O_2).[2] The oxygen transpired from the plants and floated slowly up through the lower layers of the atmosphere (troposphere) into the upper atmosphere (stratosphere), ten to thirty miles above the earth, where it was converted to ozone (O_3) by interaction with solar ultraviolet light.[3]

Over time, ozone accumulated in the stratosphere. If the ozone layer could be measured at ground-level air pressure, it would be only three millimeters thick, but at the low pressures of the thin stratospheric air it is scattered over a width of thirty-five kilometers. This gas acted as a chemical shield, keeping much of the damaging solar ultraviolet light from entering the troposphere. Because ultraviolet light kills normal living cells, no multicellular organisms (plants and animals with more than one cell) could survive before the ozone layer developed. As this protective shield thickened over time, more complex organisms could develop and evolve. The ozone layer protects the delicate web of life on earth, in much the same way that sunglasses protect delicate eyes from the damaging effects of ultraviolet light. Most life would die if the ozone layer disappeared, and the earth would return to a preevolutionary state.

In 1973, two scientists, Sherwood Rowland and Mario Molina, at the University of California at Berkeley, became worried that CFC, a man-made gas invented in 1928 by accident, could damage the ozone layer. But it was not until 1982 that a group of British scientists working in Antarctica verified that much of the ozone layer over the South Pole had disappeared. In 1983 and 1984, the American satellite Nimbus recorded very low levels of ozone, but these measurements were dismissed as inaccurate by computers that were programmed to accept only normal ozone readings.[4]

In 1987 and 1989, two planes flew high into the stratosphere and managed to measure the ozone concentration together with levels of chlorine (the element that destroys ozone). They found that 95 percent of the ozone had disappeared over Antarctica and that chlorine levels were correspondingly high. The ozone loss is apparent not just over the South Pole; since 1970, the ozone layer has decreased by 1 percent in summer and 4 percent in winter over the Northern Hemisphere in middle latitudes between latitudes 64° and 30° north. This rate of depletion was approximately twice that which computer models had predicted.[5] But satellite data released by NASA in April 1991 revealed that the ozone layer has deteriorated from 4.5 to 5 percent over the last decade, more than double the rate previously anticipated. The U.S. Environmental Protection Agency now predicts 200,000 additional deaths from skin cancer over the next decade in the United States alone.[6]

The chemicals most responsible for ozone destruction belong to a family of related compounds called chlorofluorocarbons, or CFCs. These gases are used in refrigeration, air-conditioning, spray cans, Styrofoam cups, trays, and packaging, as a plastic expander in foam furniture and car upholstery, and to clean computer chips.[7] It was originally thought that these gases were harmless, did not react with other chemicals, and were safe for human contact. However, scientists failed to foresee that some of them remain in the atmosphere for between 75 and 380 years. After release they rise up slowly over three to five years'[9] to the stratosphere, where they interact with UV radiation, which severs the chlorine atom from the CFC molecule. This chlorine atom then splits an ozone molecule (O_3) into one oxygen molecule (O_2) and one oxygen atom (O). The chlorine is then free again to continue this cycle over many years.[10] Like an environmental Pac-man, one chlorine atom can consume more than 100,000 ozone molecules over time. The level of CFCs in the stratosphere is increasing at a rate of 5 percent per year.

There are other gases that also destroy the ozone layer. The list includes carbon tetrachloride, which is used as a dry-cleaning liquid and serves as the precursor chemical for the production of CFCs; methyl chloroform, which is used for cleaning metal; HCFCs, which are substitutes for CFCs; and halons used in fire extinguishers. HCFCs have 10 to 20 percent the depleting power of CFC gases.[11]

The solid rocket fuel used in the U.S. space shuttle releases 240 tons of hydrochloric acid (HCl) into the atmosphere per launch. The chlorine atom then splits from the HCl molecule to destroy ozone molecules in the stratosphere, through which the rocket passes. If NASA continues to launch solid-fuel rocket boosters at the present rate of ten per year, some scientists predict, these alone would cause a 10 percent ozone depletion by 2005.[12] Other scientists dispute these claims, saying that the amount of chlorine added to the ozone layer from the space shuttle is quite small compared with the quantities derived from industrial chemicals globally.[13] In September 1991, NASA announced that it will shelve plans for many of the larger spacecraft in favor of smaller rockets and satellites. However, the number of launches will more than double, so the new program may still leave the ozone in great danger.[14] The United States also tests missiles for nuclear weapons delivery that use solid rocket fuel. Such tests add to the ozone depletion.[15]

The hole in the ozone layer over the South Pole is growing each year. By October 1987, 95 percent of the polar ozone had disappeared. This hole exists only between the months of September and October. It closes over for the rest of the year when stratospheric ozone in the Southern Hemisphere filters down to cover the hole. This drift mechanism depletes the total ozone over the southern regions of Australia, South America, and Africa, particularly during their summer—November to March.[16] We Australians are sun worshipers who spend much of the summer soaking up the "rays" on our wonderful beaches; despite dire warnings from dermatologists, not many people use sunscreens. Each 1 percent decrease in ozone could produce a 4 to 6 percent increase in skin cancer. If the space shuttle did indeed decrease the ozone layer by 10 percent over the next fourteen years, it could, by this calculation, increase the incidence of skin cancer in humans by 60 percent in the next ten years.[17]

MEDICAL AND BIOLOGICAL CONSEQUENCES OF OZONE DEPLETION

A small amount of UV light has always penetrated the normal ozone layer, producing the usual number of skin cancers in the past, but the incidence of skin cancer and malignant melanoma is now increasing rapidly. Some part of this increase is almost certainly due to ozone depletion. Melanoma, a dark mole that becomes malignant and is usually lethal, has doubled in frequency worldwide over the last two decades.[18] However, in Australia, where the ozone depletes each summer, the melanoma incidence has doubled in the last ten years.[19]

UV light can also damage the lens of the eye, causing cataracts and partial or complete blindness. According to the United Nations Environment Program, it has induced blindness in 12 to 15 million people and has impaired the vision of 18 to 30 million others. It also impairs the body's immune mechanism, which fights infection and cancer.[20]

Vitamin D is synthesized when UV light reacts with the skin. This vitamin is necessary for normal bone and teeth formation. A deficiency causes rickets. However, excessive UV light can induce vitamin D toxicity, with symptoms of kidney stones and abdominal and bone pains, as calcium is mobilized from the bony structure and as excessive amounts are excreted through the kidneys.

Plants, too, are very sensitive to increases in concentrations of UV light. Two-thirds of the three hundred crops tested are vulnerable, including peas, beans, melons, mustard, cabbages, tomatoes, potatoes, sugar beets, and soy beans. Forest death is very likely a result of increased levels of UV light, some trees being more susceptible than others, and ocean algae are extraordinarily sensitive to it.[21]

It is sad that we must now cover our bodies, faces, and eyes when we go out on a beautiful day. The sun, which used to be a largely beneficial presence, upon which all life depended, has become, to a certain extent, our enemy.

Because of deep scientific concern about the rapidly decreasing ozone layer, representatives of many nations met in Montreal in 1987 and signed a protocol pledging to reduce CFC production by 50 percent by the year 2000.[22] This treaty was clearly too conservative, and in June 1990 many countries agreed in London to end production and consumption of CFCs totally by 2000. But a study conducted by the U.S. Environmental Protection Agency estimates that atmospheric levels will increase more than threefold over the next hundred years, even if all CFC production ceases by the year 2000, because CFC gases will for years continue to leak from old refrigerators, air conditioners, Styrofoam cups, and plastic foam furniture.[23]

Third World countries are understandably reluctant to phase out CFCs before the majority of their people have refrigerators and the rest, and chemical companies have been reluctant to stop CFC production, claiming it will cost too much money. But we must not talk about profit and loss when the earth is dying!

There exist excellent alternatives for refrigeration gases, such as helium and zeolites. The latter is used in Sweden for refrigeration in combination with solar power or natural gas.[24] We could even go back to ammonia, the original refrigerant gas, which does not destroy the ozone.

Because political leaders in China, India, and other huge developing countries are promising their people refrigerators, it is urgent that the First World take responsibility and lead the way to alternative methods of refrigeration. Unfortunately, most industrialized countries had by April 1991 not contributed their share of $54 million in annual contributions to a UN fund to help developing countries switch to more expensive but ecologically safer chemicals.[25] If drastic steps are not taken, the ozone layer will almost certainly be depleted beyond repair. We must begin investing in companies that make safe chemicals for refrigerators. We must also be prepared not to use house or car air-conditioning and to exercise our own sweat glands during the hot weather. Legislation that bans CFC production worldwide urgently needs to be introduced. We can live without spray cans, plastic furniture, cold cars, and air-conditioning, and computer chips can be cleaned by other means. If the level of emissions of CFCs continues to grow as in the past, the ozone will be diminished by 20 percent within the lifetimes of our children. This could cause 1.5 million extra deaths from skin cancer and 5 million more cataracts in the United States alone.[26]

One last word about CFCs and other chemicals that deplete the ozone, including "substitutes" like HCFCs. They not only effectively eradicate ozone, but they are 10 to 20,000 times more efficient as a greenhouse gas than carbon dioxide, a product of the burning of fossil fuels. They are already responsible for 20 percent of the greenhouse warming, and their contribution is increasing.[27]

THE GREENHOUSE EFFECT, OR GLOBAL WARMING

The earth is heating up, and the chief culprit is a gas called carbon dioxide. Since the late nineteenth century, the content of carbon dioxide (CO_2) in the air has increased by 25 percent. Although this gas makes up less than 1 percent of the earth's atmosphere, it promises to have devastating effects on the global climate over the next twenty-five to fifty years.[28] Carbon dioxide is produced when fossil fuels—coal, oil, and natural gas—burn, when trees burn, and when organic matter decays. We also exhale carbon dioxide as a waste product from our lungs, as do all other animals. Plants, on the other hand, absorb carbon dioxide through their leaves and transpire oxygen into the air.

Carbon dioxide, along with other rare man-made gases, tends to hover in the lower atmosphere, or troposphere, covering the earth like a blanket. This layer of artificial gases behaves rather like glass in a glasshouse. It allows visible white light from the sun to enter and heat up the interior, but the resultant heat or infrared radiation cannot pass back through the glass or blanket of terrestrial gases. Thus the glasshouse and the earth heat up.

In one year, 1988, humankind added 5.66 billion tons of carbon to the atmosphere by the burning of fossil fuels, and another 1 to 2 billion tons by deforestation and the burning of trees. Each ton of carbon produces 3.7 tons of carbon dioxide.[29]

But carbon dioxide accounts for only half of the greenhouse effect. Other gases, the so-called trace gases, which are present in minute concentrations, are much more efficient heat trappers.[30] As was previously mentioned, CFCs are 10 to 20,000 times more efficient than carbon dioxide. Methane is also very efficient (20 times more effective than carbon dioxide) and is released at the rate of 100 liters per day from the intestine of a single cow. For example, Australia's cows make an annual contribution to global heating equivalent to the burning of thirteen million tons of black coal (about half the coal used in Australia per year). The scientists Ralph Laby and Ruth Ellis, from the Australian Commonwealth Institute and Research Organisation, have developed a slow-release capsule that diminishes by 20 percent the production of methane by bacteria in the rumen of cows. (Methane is also a wonderful gas for heating and lighting houses; for example, Laby and Ellis estimated that two cows produce enough methane to heat and light an average house!)[31] Further sources of methane are garbage dumps, rice paddies, and termites. Nitrous oxide is another greenhouse gas, a component of car and power plant exhausts, of chemical nitrogenous fertilizers, and of bacterial action in heated, denuded soil. Nitrous oxide has increased by 19 percent over preindustrial levels and methane by 100 percent.[32] A report from the World Wide Fund for Nature published in August 1991 stated that carbon dioxide emissions from aircraft flying at altitudes of ten to twelve kilometers account for 1.3 percent of the global warming. However, the nitrous oxide that aircraft also emit is an extremely efficient heat trapper at that height and may increase global warming by 5 to 40 percent.[33]

Within fifty years, the "effective carbon dioxide concentration" (CO_2 and trace gases) will probably be twice that of preindustrial levels, raising global temperatures 1.5° to 5.5°C (2.7° to 10°F).[34] Because many scientific variables—heat trapping by clouds, change in radiation over melting ice caps, and so on—are not well understood, this rise in temperature could be as high as 10°C (18°F). Other scientists say the earth could cool several degrees. But all agree that we are in trouble.[35]

Such a rapid change in climatic conditions has never occurred in human history. If global heating were at the lower predicted level, it would match the 5°C warming associated with the end of the last ice age, 18,000 years ago. But this change would take place ten to a hundred times faster.[36] And at present temperatures, a 5°C increase would cause global temperatures to be higher than at any other time during the last 2 million years.[37]

What will happen to the earth? Let us look at a worst-case scenario. Changes of climate could have devastating consequences in the tropical forests and food-growing areas of the world, causing extinction of many plant and animal species over a few years, in evolutionary terms. Dust bowls could develop in the wheat belt of the United States, creating a situation like that described in *The Grapes of Wrath,* and the productive corn and wheat belt might migrate north into Canada and into the Soviet Union.[38]

Already the futures markets are speculating that productive banana and pineapple plantations will develop in the middle of arid Australia. Cyclones, tidal waves, and floods will almost certainly affect temperate areas of the world, which were previously immune to such catastrophes.[39]

Sea levels will probably rise as the warming oceans expand, and great areas of land will be flooded, particularly during storms. Rivers, lakes, and estuaries will have their courses and boundaries changed forever. This will disturb the hatching habitats of millions of fish.[40]

Because about one-third of the human population lives within sixty kilometers of the sea, millions, or even billions, of people will either be killed by floods or storms or be forced to migrate to higher levels, thereby severely dislocating other urban and rural populations. These refugees will create chaos as they move into established rural areas, towns, and cities. Food production will already have been disrupted by the change in climate, and a redistribution of the scarce remaining resources will probably not happen.[41]

As sea levels rise, beautiful cities, including Venice and Leningrad, will be submerged, and even Westminster Abbey and the houses of Parliament, in London, will be threatened. Many beautiful, exotic Pacific islands will be underwater. Sea levels could rise seven feet (2.2 meters) by the year 2100, according to the U.S. Environmental Protection Agency.[42]

It is possible that the polar ice caps will melt; alternatively, the Antarctic snow cover might increase in volume as warm air induces a buildup of snow-forming clouds over the South Pole.[43] (Warm air promotes the evaporation of water from the earth's surface, thereby thickening the cloud cover.)

The aquatic food chain will be threatened because the base of the pyramid of the ocean food chain—algae and plankton—will be seriously affected. These ubiquitous single-celled plants are food for primitive life forms and are themselves consumed by more evolved species of fish. Some forms of algae and plankton will be threatened by rising sea temperatures, and many are extremely sensitive to UV light—so much so that some species have a built-in mechanism enabling them to dive from the surface to lower depths at midday, when the UV light is most powerful, to escape the lethal radiation. Therefore, as the temperature rises and as the ozone diminishes, this essential element of the food chain will be jeopardized.

Moreover, plankton and algae, together with trees and plants, are nature's biological traps for elemental carbon from atmospheric carbon dioxide, 41 percent being trapped in sea plants and 59 percent in land plants. Higher concentrations of atmospheric carbon dioxide will promote the growth of algae. But if algae are threatened by global warming and ozone depletion, this hypothetical fertilizer effect will become irrelevant. By increasing the atmospheric concentration of carbon dioxide from man-made sources, we are thus also threatening the survival of trees, plants, algae, and plankton.

Forests, too, are terribly vulnerable to climatic change and ozone destruction. Because temperature changes will be relatively sudden, specific tree species will not have thousands of years to migrate to latitudes better suited to their survival, as they did at the end of the last ice age. When the ice cap slowly retreated northward, the spruce and fir forests moved from the area of the United States into Canada at the rate of one kilometer per year. Although some plants that adapt rapidly will thrive under changed circumstances, most forests will die, and along with them many animal and bird species.[44]

Interestingly, although sudden global warming will kill large numbers of trees, increased carbon dioxide concentrations will actually stimulate the growth of those that remain, because the gas is a plant food during photosynthesis and thus acts as a fertilizer.[45] Therefore, as forests become extinct in the unusually hot climate, some food crops and surviving trees will grow bigger and taller. Unfortunately, many weeds are even more responsive to high carbon dioxide levels than crop plants are, and they will almost certainly create adverse competition.[46]

Another factor to consider in this rather dire biological scenario is that faster-growing crops utilize more soil nutrients. Hence more artificial fertilizer will be needed, and, since electricity is required for its production, more carbon dioxide will be added to the air. But nitrogen-containing fertilizers themselves release the greenhouse gas nitrous oxide into the air.[47] In addition, as soil heats, vegetable matter decays faster, releasing more carbon dioxide. These are just a few of the interdependent and variable effects of global warming that are so difficult to calculate.

When forests are destroyed by greenhouse and ozone deforestation, or by chainsaw and bulldozer deforestation, massive quantities of rich topsoil will be lost forever as floods and erosion wash it out to sea. Downstream waterways will overflow their banks as rain pours off the denuded high ground, and when the floods subside, the once deep rivers will be silted up from the eroded topsoil. Large dams designed for predictable rainfalls could collapse and drown downstream populations, and associated hydroelectric facilities would then be destroyed.

Decreased rainfall in other parts of the world will reduce stream runoff. For example, a rise of several degrees Celsius could deplete water levels in the Colorado River, causing severe distress for all communities that depend upon the river for irrigation, gardening, drinking water, and so forth. The water quality will also suffer, because decreased volumes will not adequately dilute toxic wastes, urban runoff, and sewage from towns and industry.[48] Until April 1991, when rain began to fall again in some quantity, California experienced a severe five-year drought, whose impact was rapidly becoming critical. After this April rainfall, the California drought continued unabated. That may be an omen of worse to come.

Cities will be like heat traps. For instance, Washington, D.C., at present suffers one day per year over 38°C and thirty-five days over 32°C (100°F and 90°F). By the year 2050, these days could number twelve and eighty-five, respectively.[49] In that case, many very young and many old and infirm persons would die from heat stress, and there would be a general temptation to turn on air conditioners, which, of course, use CFCs and electricity, whose generation produces more carbon dioxide. People will thus be in a catch-22 situation—damned if they do and damned if they don't.

2

Atmospheric Degradation: Causes and Some Solutions

How did the problem of atmospheric degradation become so alarming, and what are the solutions?

When CFC was first concocted, in 1928, nobody understood the complexities of atmospheric chemistry, and during subsequent decades scientists really believed that chlorofluorocarbons were ideal for refrigeration, air conditioners, plastic expanders, spray cans, and cleaners for silicon chips.[1] Industry became so heavily invested in its production that it now finds it very difficult to cut back, even though the environmental consequences of not doing so will be severe.

At the 1990 London International Conference on Ozone, a group of enthusiastic young Australians made representations begging for a safe future. One middle-aged conference participant approached the teenagers and said, "Look, my attitude is that if you are on the *Titanic,* you may as well have the best berth." Such cynical acceptance of the planet's demise is suicidal and demonstrates a total lack of a sense of responsibility for the next generation. In fact, since the release of the 1987 Montreal protocol to reduce CFC production, some major chemical manufacturers have attempted to deny that these chemicals destroy ozone in the atmosphere. At international meetings, there is continual intense lobbying to protect the interests of industry.[2]

In the early years of the Industrial Revolution, no person could have predicted the atmospheric havoc that the internal-combustion engine and coal-fired plants would wreak. Even during the 1930s and 1940s, when General Motors, Standard Oil, Phillips Petroleum, Firestone Tire and Rubber, and Mack Manufacturing (the big-truck maker) bought up and destroyed the excellent mass transit systems of Los Angeles, San Francisco, and most other large U.S. cities in order to induce total societal dependence on the automobile, global warming was a vague future threat.[3] These companies were subsequently indicted and convicted of violating the Sherman Antitrust Act.

But now that we understand the coming disaster, we are in a position to act. In order to act, we must be willing to face several unpleasant facts.

FACT NUMBER ONE. The United States, constituting only 5 percent of the earth's population, is responsible for 25 percent of the world's output of carbon dioxide.[4] It uses 35 percent of the world's transport energy, and an average-size tank of gasoline produces between 300 and 400 pounds of carbon dioxide when burned.[5] Together, the United States and the Soviet Union consume 44 percent of the world's commercial energy.[6]

In China, by contrast, there are 300 million bicycles, and only one person in 74,000 owns a car. Each year three times more bicycles than cars are produced. Domestic bicycle sales in 1987 came to 37 million—more than all the cars bought worldwide.[7] Motor vehicles globally produce one-quarter of the world's carbon dioxide, and in the United States transportation (cars, buses, trains, and trams) produces 30 percent of all the carbon dioxide. Transport consumes about one-third of all the energy consumed globally. The United States also produces 70 percent of the carbon monoxide gas (which leads to deoxygenation of the human blood), 45 percent of the nitrous oxides (which cause acid rain), and 34 percent of the hydrocarbon chemicals (many of which are carcinogenic).[8]

In 1985, there were 500 million motor vehicles in the world, 400 million of them cars. Europeans and North Americans owned one-third of these.[9]

FACT NUMBER TWO. In order to reduce carbon dioxide production, cars must be made extremely fuel efficient, and some computer models and prototype automobiles can indeed achieve 60 to 120 miles per gallon (mpg) by means of lightweight materials and better design.[10] But these techniques are not being employed. In 1987, U.S. car manufacturers dropped most of their research on fuel-efficient cars, and in 1986 the fuel-efficient standard, or minimum mpg, in the States was only 26 mpg.[11] In 1991, it was still only 27.5 mpg, and the Bush administration has resisted any move to increase fuel efficiency in cars.[12] In fact, the president's new energy plan of 1991 barely deals with these issues, and does not deal at all with mass transportation. It is more efficient to transport hundreds of bodies in one train than hundreds of single bodies in hundreds of cars. Furthermore, the construction of sleek state-of-the-art trains would constructively reemploy the one in eight people in California who currently are employed producing weapons of mass destruction.[13] Far more people will work in this wonderful new civilian industry than in the obsolete weapons industry, because the military sector is capital intensive, whereas the civilian sector is labor intensive. The corporation that first accepts this challenge could become the world's leading producer of global mass transit systems and could earn huge profits while saving the planet.

Cars can be fueled with solar energy. In 1990, an international solar car race across Australia was held. The cars achieved the acceptable speed of approximately 60 mph. They were slow to accelerate, but who needs cars that go from 0 to 90 mph in a matter of seconds? Cars can also be fueled with natural gas, which generates less carbon dioxide than gasoline does, and with alcohol. By investing heavily in this form of energy, Brazil has been helped to become somewhat energy independent. In 1988, alcohol provided 62 percent of Brazil's automotive fuel. Although alcohol is relatively expensive, it gives off 63 percent less carbon emission than gasoline. This excellent form of energy production is renewable, and marginal land can be used to grow crops that can be converted into alcohol. The United States already produces twenty million barrels of alcohol by the fermentation of corn. As oil prices climb, alcohol will obviously become a viable fuel alternative.[14]

Cars can also be fueled with hydrogen; the technology is available. One can drink the exhaust of a car powered by hydrogen, because when hydrogen burns it produces pure water. Unfortunately, major U.S. auto companies seem resistant to being inventive and creative. They stick to old, outdated designs and have even resorted to copying the latest Japanese designs—a rather sorry setback for an industry that once led the world in automobile technology. It could overtake the Japanese industry by manufacturing solar-, hydrogen-, and alcohol-powered cars, while helping to save the planet.

FACT NUMBER THREE. Bicycles use human energy and save global energy. They are clean, efficient, and healthful for human bodies. Roads must give way to bicycle tracks. In China, special bicycle avenues with five or six lanes are separated from motorized traffic and pedestrians.[15] This sort of planning is required by a large percentage of the population of the United States and of the Western world. The arrangement is simple, easy, cheap, and clean. Distant, large-scale supermarkets and shopping malls accessible only by car will become obsolete as people demand small, convenient shops within walking distance of their homes. We can reestablish small community shopping centers, where people meet each other and socialize and where the emphasis is on the community rather than on consumerism. What a healthy, exciting prospect!

FACT NUMBER FOUR. Buildings can be made extremely energy efficient. Improved designs for stoves, refrigerators, and electric hot-water heaters can increase energy efficiency by between 5 and 87 percent. The sealing of air leaks in houses can cut annual fuel bills by 30 percent, and double-paned insulated windows greatly reduce energy loss. Superinsulated houses can be heated for one-tenth of the average cost of heating a conventional home. In the United States, 20 percent of the electricity generated is used for lighting, but new fluorescent globes are 75 percent more efficient than conventional globes. Theoretically, then, the country could reduce its electricity usage for lighting to 5 percent of the total.[16] This, together with other conservation measures, would cause the closing down and mothballing of all nuclear reactors in the States, because 20 percent of all the electricity used there is generated by nuclear power.[17]

I lived in the beautiful city of Boston for fourteen years and grew to love the old New England houses. But now that I am more aware of the fate of the earth, I realize that these are totally inappropriate dwellings. They are big, rambling, leaky, and inefficient. The vast quantities of oil required to heat these large volumes of enclosed air through a long Boston winter adds to carbon dioxide greenhouse warming. When these handsome houses were built, in the last century, no one imagined that the earth would someday be in jeopardy. Fuel supplies seemed endless, and the air was relatively clean.

Houses of that kind can be made somewhat fuel efficient by the insulation of walls, ceilings, and windows and by the sealing of all leaks. To encourage such reform, the federal government must legislate adequate tax incentives, for in the long run these will provide insurance for our children's future.

Actually, solar buildings are now in an advanced stage of design and development. The need is for legislation that requires all new buildings, residential and office, to be solar designed—with large heat-trapping windows oriented toward the south and with floors made of tiles and cement, which trap the sun's heat during the day, and appropriate window insulation, which retains the heat at night. Solar hot-water panels and solar electricity generation are relatively cheap and state-of-the-art. Firms that manufactured such equipment would make large profits. Householders would benefit because they would become independent of the utilities; they could even sell back electricity to the local utility at off-peak hours. Indeed, some Americans are already vendors of electricity.

Solar technology would then become highly efficient and cheap, and a huge market would open up in the Third World. The industrialized countries could assist billions of people to bypass the fossil-fuel era, so they could generate electricity from solar and wind power and use solar cookers and solar hot-water generators. This is a signal solution to the problem of ongoing global warming. The First World must help the Third World bypass the fossil fuel era if the earth is to survive.

Attention should also be given to the strange high-rise buildings covered in tinted glass that seem to be in vogue in many U.S. cities. These are not solar buildings. The windows cannot be opened to allow ventilation during the summer, and they must be cooled with air conditioners, which use ozone-destroying CFCs and carbon dioxide-producing electricity. In the winter, heat leaks from the windows like water through a sieve. And these buildings are generally lit up like Christmas trees at night, for no apparent purpose, by energy-inefficient lighting. Dallas, Houston, and Los Angeles boast numerous of these monstrosities, many of which now sit empty and idle, built by speculators who cashed in on the savings and loan scandal. (I used to wonder as I traveled through the United States in the 1980s, why the Sun Belt was thriving. Now we know! This prosperity was a by-product of the deregulation of the savings and loan industry by the Reagan administration.)

We all must become acutely conscious of the way we live. Every time we turn on a switch to light a room, power a hair dryer, or toast a piece of bread, we are adding to global warming. We should never have more than one light bulb burning at night in our house unless there are two people in the house in different rooms—then two bulbs. Lights must not be left on overnight in houses or gardens for show, and all lights must be extinguished in office buildings at night.

Clothes dryers are ubiquitous and unnecessary. In Australia, we dry our clothes outside in the sun, hung by pegs from a line. Americans can do the same in the summer, and in the colder climes, like Boston's, clothes can be hung on lines in the cellars in the winter. In some American cities there are laws prohibiting people from hanging clothes on lines outside, on grounds that it is not aesthetically pleasing. This method of drying offers, in fact, an easy and efficient step toward the reduction of atmospheric carbon dioxide and radioactive waste. Clothes dryers use over 10 percent of the electricity generated in the States,[18] a large fraction of that generated by nuclear power. And bear in mind that electrically operated doors, escalators, and elevators all contribute to global warming.

MONOPOLISTIC CONTROL OF ENERGY PRODUCTION

We must examine the monopolistic control of energy production to understand why the American public has been led to believe that energy consumption should be on an ever upward curve, and why the federal government has not encouraged the use of solar systems and tax incentives for energy conservation.

The oil companies are among the most powerful corporations in the world. Many are also involved in the production of nuclear power and nuclear weapons by their mining of uranium. The Department of Energy (DOE), which should be concerned only with energy production, actually oversees the manufacture of nuclear weapons and has supervised the production of 35,000 hydrogen and atomic bombs since 1945—enough to "overkill" every Russian person forty times. (The United States, like a person addicted to alcohol, was in 1991 still manufacturing delivery systems for nuclear weapons, despite the end of the cold war. Military contractors making nuclear weapons and delivery systems enjoy profits of up to 75 percent per year, while civilian industries show profits of only 15 percent. Incredibly, about fifty-five cents of every federal tax dollar paid by the American people fund weapons production—for a nonexistent enemy.)[19]

The details of the business of energy production have been jealously guarded by the utilities, oil and nuclear companies, and government departments, which often work together secretly.[20]

As the *New York Times* reported on November 6, 1989, "The Energy Department's reliance on contractors and consultants for basic Government functions is pervasive but largely outside public scrutiny. . . . The department has increasingly retained consultants to help perform 'virtually all basic functions' ". The situation is so bad that in 1989 the secretary of energy, Admiral James D. Watkins, was extremely embarrassed to discover that a testimony he gave to Congress had been written by the very corporations that are supposed to be regulated by the DOE.[21]

So the energy policies of the DOE are determined by the vested interests of its contractors. Despite dire predictions of global warming, much emphasis is placed on coal and oil and nuclear power, but virtually none on alternative energy techniques. George Bush's energy plan of 1991 increased the funding for solar power and the like from zero (set by Ronald Reagan) to 5 percent, while increasing the funding for nuclear power by 40 percent, despite the disasters at Three Mile Island and Chernobyl. It called for the building of several hundred nuclear reactors within the next decades at a cost of $390 billion to over $1,305 billion.[22] You see, the oil and nuclear companies have a hidden agenda. They figure that if they can scare the general public enough with the threat of impending global warming, they can persuade it to accept nuclear power once again. (Yet so strong is public sentiment against nuclear energy that all orders for new nuclear reactors have been canceled since 1974.)

Immediately following the Persian Gulf war, full-page ads appeared in *Time* and *Newsweek,* depicting the scary faces of Muammar Qaddafi and Saddam Hussein, Ayatollah Khomeini, and President Hashemi Rafsanjani, the new leader of Iran, asking whether the American people wished to be hostage to these men for their energy. These ads, paid for by the public relations wing of the nuclear power industry, contain several dangerous lies. (1) Oil is used not to generate electricity but for transportation—only 1.5 percent of all imported oil goes to the generation of electricity.[23] (2) The ads advance this lie to encourage public acceptance of nuclear power. (3) Huge quantities of carbon dioxide are generated during the manufacture of nuclear reactors and the uranium fuel and the dismantling of twenty-year-old obsolete reactors. (4) The ads give only two choices for electricity generation, oil or nuclear; they do not mention coal, or, more importantly, alternative energy sources. (The medical consequences of nuclear power will be described in a later chapter.)

To add insult to injury, George Bush, who made his fortune from oil, has shown a persistent reluctance to join the community of nations, many of which are agitating to lower global carbon dioxide production. He has refused to lower it in the United States over the next ten years.[24] Some U.S. corporations have also lobbied intensively against two congressional bills designed to reduce carbon emissions by 20 percent over the next decade.[25]

The priorities in regard to energy production are so crucial that the future of the world hangs in the balance, depending upon the decisions we make now, in the early 1990s. I therefore suggest, as a matter of urgency, that we split the bureaucracy of the Department of Energy into three departments:

1. Department of Nuclear Weapons, whose one, urgent task is to begin dismantling or decommissioning hydrogen bombs.
2. Department of Radioactive Waste, whose task is to enlist large numbers of the world's best scientists to work on the almost insurmountable problems of safe long-term storage of radioactive wastes, to clean up contaminated radioactive bomb factories, and to develop relatively safe methods of decommissioning nuclear reactors at the end of their twenty- to thirty-year lifespan. (The medical problems associated with radioactive waste are described in chapter 4.)
3. Department of Energy, whose task is to supervise and develop renewable and safe energy sources for the next decade and thereafter.

Renewable energy sources (wind, solar, geothermal, and so on) could theoretically provide a total energy output equal to the current global energy consumption. Today these sources already provide approximately 21 percent of the energy consumed worldwide and are freely available to be developed further.[26]

Solar power will soon yield electricity as cheap as coal-fired electricity. In fact, scientists at the U.S. Solar Energy Research Institute estimate that photovoltaic solar systems could supply over half the U.S. electricity within forty to fifty years. This technology will decrease in price as it is mass produced, modified, refined, and made more efficient. Solar water and household heating is already widespread in Australia, Greece, and the Middle East.[27]

Wind power offers an obvious and benign technique that is being used to generate electricity in many countries, including Greece, China, Australia, Israel, Belgium, Italy, Germany, Britain, the United States, and Denmark. Since 1974, fifty thousand wind machines have been built, mainly in California and Denmark. Wind "farms" cover areas of the desert between Los Angeles and Palm Springs, and by the year 2030 wind power could provide 10 percent of the world's energy.[28]

Geothermal power, which taps into the intrinsic heat and lava trapped below the earth's crust, is already being used to good advantage in New Zealand, Iceland, and Hawaii, and there is much potential for its use in the U.S., Soviet Union, and Central America. Output is increasing by 15 percent per year.[29]

Tidal power utilizes the twice daily changes in sea levels to generate electricity. It is suitable only for certain coastlines, but it certainly offers great possibilities in places where the tides vary twenty to a hundred feet per day. Wave power is another dynamic area awaiting development.

Hydropower has been expanding by 4 percent annually worldwide, and the potential for further expansion is vast.[30] Hydroelectric and geothermal power provides over 21 percent of the world's electricity.[31] Electricity generated at dams and waterfalls crosses borders and can be used in other countries; for instance, New England uses Canadian hydroelectricity. Hydroelectric dams that flood large areas of natural forests are ecologically dangerous, and careful planning is essential before and during their construction.

Cogeneration is a wonderful method for harnessing heat usually wasted in factories. One technique, used extensively in the Soviet Union and in Scandinavia, is to heat water and pipe it to warm whole towns and cities. Another is to use waste steam to drive electricity-generating turbines, to run refrigerators, and to power industrial machinery. An ordinary power plant is 32 percent efficient, but a cogenerator consuming the same amount of fuel is 80 percent efficient.[32]

Conservation can save large quantities of energy. Society must invest in highly efficient light bulbs, refrigerators, stoves, cars, and street lighting. Energy-efficient equipment uses one-third to one-half less energy than does conventional technology. Much of it has already been invented, but monopolistic corporations tend to encourage distribution of inefficient equipment, thus leading to increased electricity consumption. For example, General Electric manufactures not only nuclear reactors but also hair dryers, toasters, stoves, and refrigerators. Is it not therefore in GE's best interests to encourage people to use more electricity with less efficient appliances and to use electric brooms, electric hedge clippers, and electric lawn mowers instead of ones operated by muscle power?

But energy-efficient investments are much cheaper financially and ecologically than the building and operating of coal or nuclear plants. Patents for wonderful energy-saving inventions abound, but most inventors lack the money to develop their product. And corporations seem uninterested in pursuing or financing such inventions.

Not least, utilities hide enormous government subsidies that they receive for fossil fuels and nuclear power. This deception makes renewable energy appear to be more expensive. Because utilities enjoy an almost total monopoly in energy advertising and technologies of energy production, it is very hard to understand and dissect their propaganda. Solar-heating systems and photovoltaic cells endow people with energy self-reliance, but clearly such self-sufficiency is not and will not be seen to be in the best interests of the utilities.

Trees and other plants (biomass) are sources of energy mainly in the developing world. In India, people even burn pats of cow dung for cooking. But inhabitants of these countries often decimate their forests for short-term survival. I have seen Indian women spend a whole day walking to a patch of trees, gather the wood, and walk home for another day in order to cook food for their families. The burning of wood adds to atmospheric carbon dioxide.

Deforestation is leading to desertification in many countries—to creeping deserts and utter destruction of the land. Brazil is even using parts of the Amazon forest to fuel iron ore smelters.[33] So wood is not necessarily a good fuel and needs to be replaced by solar, wind, and other kinds of power. Still, garbage and agricultural wastes can be used to produce methane, an excellent gas for cooking and heating. Biomass supplies 12 percent of the energy worldwide and up to 50 percent in some poor countries.[34]

Industrial efficiency has been shown to have enormous potential. It must be developed on a massive scale, for industry uses 40 to 60 percent of the available energy in the developed countries and 10 to 40 percent in the developing countries.[35]

A CARBON TAX

If all fossil fuels were taxed to avoid climate change, the ecosphere could be brought into a relatively stable equilibrium. In the United States, this tax would raise the price of electricity by 28 percent and that of a gallon of gasoline by seventeen cents, but it would produce $60 billion in revenue, and this money could then be earmarked for alternative-energy facilities and conservation. In India, the tax would raise $17.5 billion.[36] The international community within the United Nations must endorse this tax proposal. According to the Worldwatch Institute report of 1990, in order to stabilize atmospheric greenhouse gas concentrations by 2050, net carbon emissions will need to be reduced by two billion tons per year. So, given a probable global population of eight billion by then, all people will require levels of net carbon emissions similar to India's today, which is only one-eighth of the current levels in Western Europe. Furthermore, 20 percent of the global carbon tax could be diverted to Third World reforestation, benign energy production, and renewable energy sources.[37]

A 12 percent reduction in global greenhouse gas emissions by 2000 seems an appropriate interim goal if we are to achieve a stabilization of carbon dioxide concentrations by 2050. This means that the United States and the Soviet Union would have to reduce carbon dioxide production by 35 percent over the next ten years. To be fair, though, Kenya and India could actually increase carbon dioxide emissions, because they produce so little at present. If we fail to make these important decisions and if the industrial countries maintain present-day emission levels, the Third World, by emulating the First World, could increase the quantity of carbon dioxide by some 20 to 30 percent by the year 2000 and by 50 to 70 percent by 2010, as its fuel use and population base expand.[38]

I don't think we have any choice in these matters, and the sooner we knuckle down to the task, the sooner we can reassure our children that they will inherit a viable future.

3

Trees: The Lungs of the Earth

A tree is a noble organism, unique in its beauty and home and refuge for birds, insects, and small animals. The eminent biologist Edward Wilson found on one tree in Peru forty-three ant species, belonging to twenty-six genera, which is approximately equivalent to the diversity of ants in the United Kingdom.[1]

I live in a land—Australia—where gum or eucalyptus trees predominate. There are more than six hundred varieties, and even though many are similar in appearance, each tree has a design different from that of any other. When I was growing up, I thought that gum trees were boring and all the same. It was not until I had lived in the Northern Hemisphere for fourteen years that I began to really appreciate the unmatched flora and fauna of Australia. Australians tend to take their natural heritage for granted—familiarity breeds contempt.

When the English invaded Australia some two hundred years ago, they destroyed large areas of native bush in order to plant rosebushes and deciduous Northern Hemisphere trees and to create rolling meadows reminiscent of England's. They clearly felt uncomfortable with odd-shaped animals like kangaroos, screaming colorful parrots, and strange prickly bushes and trees. Originally, 14 percent of this vast desert continent was forested; now only 7 percent is. We have induced the extinction of 100 species of plants, 18 mammals, 3 birds, and 1 reptile; and 4,000 other species are on the endangered list. We have cleared six million hectares of rain forest, and Australia has three times the area of degraded land per capita than do comparable countries.[2]

The trees were felled and the land cleared for agriculture. But over the last twenty years, the government has encouraged aggressive deforestation and land degradation by inviting Japanese corporations to cut down our last remaining forests. The wood is fed into a mulcher, converted to wood chips, and shipped to Japan to be manufactured into computer paper. So alarming is the rate of deforestation that, if present trends continue, the koala may become extinct as its habitat contracts and disappears. During my electoral campaign for the federal Parliament in 1990, I was told about a new development adjacent to a main highway. The real estate agent decided to fell a stand of magnificent gum trees next to the road, and the following day koalas and echidnas were staggering across the freeway and being hit by speeding cars.

You see, in terms of the biology of the planet, *development* is a euphemism for *destruction*. Even the frequently used term *sustainable development* involves an exercise in confusion. In a world where all resources are finite—forests, minerals, soil, air, and water—continued use and abuse of them can have only one end: the depletion and destruction of most life. Once a forest that has taken thousands of years to evolve into a system of complex biodiversity is destroyed, it takes hundreds of years to regenerate.

Trees are more than just havens for animals, birds, insects, and humans; they are also the lungs of the earth. Just as we breathe oxygen into our lungs and exhale carbon dioxide, so trees breathe carbon dioxide into their leaves and exhale oxygen. Trees are really upside-down lungs: their trunks are equivalent to the trachea, their branches to the right and left main bronchi, and all their branching twigs and leaves to small bronchi and alveoli, or air sacs, where the exchange of oxygen and carbon dioxide takes place. Tree trunks and branches may appear solid, but they are really rigid channels that transmit water and nutrients to the leaves, the way the trachea and air passages transmit air to the alveoli.

Trees are therefore an organic necessity to the biological health of the planet. As human beings fill the air with carbon dioxide and destroy the ozone layer with man-made chemicals, trees offer an excellent means of buffering these effects. It has been calculated that if an area the size of Australia or the United States were planted with trees, the air could be cleared of carbon dioxide released from fossil fuels.[3]

Just as the planting of trees replenishes the atmosphere, so deforestation helps destroy it. Because a tree spends two hundred years absorbing carbon dioxide and storing the carbon in its wood, when we chop it down and burn it either as wood or as paper, we release two hundred years of trapped carbon as carbon dioxide, thus exacerbating the greenhouse effect. We thus need to plant trees and to stop felling them, for every tree is precious, in Australia, in the Amazon, or in Washington State. The ozone layer and greenhouse gases do not recognize national boundaries, so every tree felled has global ramifications.

Once upon a time, the countries of the Middle East were covered with a humid tropical rain forest. Over time, the trees were chopped down so that wooden boats could be constructed and civilization developed, and now these countries are virtual deserts. In fact, most countries in the Northern Hemisphere were covered with forests that teemed with life; now the trees and the wildlife are almost gone. When forests vanish, the climate tends to change, making natural reforestation almost impossible. In some arid climates like Israel's, intensive drip irrigation has been used to initiate reforestation programs. But it will be many years before these trees reach such a mass that they will affect the climate.

In November 1989, I went to Puerto Ayacucho, in Venezuela, hired a dugout canoe, and set off down the Orinoco River, which is a tributary of the Great Amazon River. All my life I had dreamed of the Amazon forest, and as I read of its impending destruction, I knew I had to see and experience it while it still stood in its magnificence. I took my twenty-four-year-old son. Our boat, manned by a driver, a cook, and a guide, was at least fifty feet long, made from a single tree. We slept in hammocks located on the prow and spent ten days on the river. The climate was hot and humid, almost unbearable when we pulled in at huge sandstone rocks jutting out into the river to cook our meals, but simply beautiful as we moved slowly along the river and the soft perfumed breeze slid past our faces. I have never smelled such clean, pure, scented air in all my life. We would wake up early and embark on the day's journey at sunrise. William and I lay rocking in our hammocks, gliding into the rosy pink sky surrounded by the most magnificent forest, filled with exotic trees I had never seen before. We encountered hundreds of different palms, huge flowering trees covered with pink, white, or yellow blossoms, and colorful birds screaming as they flew from the edge of the river deep into the jungle. The undergrowth was so thick and matted as to be impenetrable. At night, white freshwater dolphins snorted and spurted in the water, while the jungle was alive with raucous screams of monkeys and other animals. As we lay under our mosquito nets, the sky was ablaze with the brightest stars I had ever seen.

When we stopped the boat, we were immediately covered with biting insects of all sizes and shapes. The bees were ten times larger than any I had ever observed before, and tiny midges, though hardly visible, left a subcutaneous hemorrhage and itch that lasted six weeks. The bites itched so much that it took great willpower not to scratch, and despite the application of the most carcinogenic insect repellent used in Vietnam, the insects were not deterred. I decided that jungles are not for people but for insects.

The jungle went on forever—an ocean of trees, covering an area the size of the United States. It seemed to me that humans could never have a significant impact on this vast creation of nature. But then I remembered the Middle East and the Sahara

Desert and realized that, in those days, it took two men nearly a week to fell a single tree. Now the chainsaw, bulldozer, and match are much more efficient.

The mornings were always sunny and intensely beautiful. Every afternoon at about three, beautiful billowy cumulus clouds would tower above the river and unleash a thunderstorm and deluge that would last one to two hours. Every day, each tree transpires into the air hundreds of gallons of water, which evaporates and creates the afternoon rain. When the forest is destroyed, the transpiration of water ceases, the rain stops, the soil dries out, and the region becomes a desert. It is said that if half the Amazon jungle is cleared, the existence of the remaining forest will be threatened by decreased rainfall.[4] The Amazon river system is the world's largest, and the delta of the Amazon is 16.8 miles wide, six times the size of the English Channel at Dover. So great is the outflow of fresh water that fifty miles out to sea you can still drink fresh water from the ocean. The river is surrounded by half the world's rain forest.[5] The Amazon seemed to me a perfect example of the microclimates created by forests, and of the change of climate that ensues when man destroys the cathedrals of nature.

As we started our journey, we noticed clearings along the river populated by primitive buildings adorned with crosses. These were Indian villages. We stopped at one of them, and the chief came out to greet us, so drunk that he could hardly stand. I realized with a sense of shock that these Yanomami Indians had been "civilized" by Catholic missionaries and taught to adopt a Western life-style. They wore skirts, or T-shirts and trousers, and looked uncomfortable in them. In the past, many had been infected with our diseases such as malaria, venereal disease, tuberculosis, influenza, and measles and had died. Others had become addicted to our drugs, and all had lost their innate ability to live with and from the forest—to eat forest foods—and they existed, to a large degree, on Western diets of flour, sugar, canned foods, and so on.

As we traveled deeper into the forest, these villages become less common, until eventually we met a canoe full of naked men sitting proud and tall in their boat, armed with spears and bows and arrows, painted with ocher, and adorned with feathers. They were off for a day's hunting. A little farther down the river, we came upon a group of women and children on the edge of the water. There was no clearing, but we saw tenuous tracks through the matted jungle. Some of the younger children had swollen bellies; the older children and women, by contrast, were not only healthy but were the most curious, alive, vibrant people I had ever seen. They climbed all over the boat, and one of the women pointed to my brassiere, which was drying on the side of the canoe, and indicated that she would like to try it on. I reluctantly agreed, and within five minutes, twenty women had squeezed themselves, one after another, into this garment before handing it on to others. It turned from white to dark brown, and as we departed I decided, with some misgivings, to leave it with them, since they liked it so much. One more token of Western civilization to pollute their way of life!

The more I saw of these wonderful indigenous people, the less I thought our culture was at all civilized. They live in harmony and peace with the forest, protecting and respecting it, while we rape and destroy it for "economic" reasons.

The Indians have for thousands of years practiced a form of cultivation in the jungle that is, of course, ecologically sound. They fell a group of trees and allow them to fall outward radially. They then burn the leaves on the outer circle to fertilize the soil and plant indigenous plants and fruits in a radial pattern, the plants that need more fertilizer being in the outer circle. Eventually, the forest encroaches on the garden, and within ten to fifteen years the cultivated circle reverts back to the forest.[6]

We met a Catholic priest in a small Brazilian village who had immigrated from Ireland thirty years earlier. He told me that the priests were doing wonderful work, reaching out and finding remote, untouched Indian tribes deep in the jungle and converting them to Christianity. Yet the damage that the white people have done to indigenous cultures over hundreds of years is irreparable, and much of it has been done, I fear, in the name and for the benefit of religion.

Indians have coexisted with the Amazon for ten thousand years. But now the Yanomami Indians are being massacred by local governments, by miners, and by the army, because they are regarded as an obstacle to development. And although the Catholic church may have instigated the destruction of their precious culture, in this respect, it is one of their few defenders.[7] Brazilian laws recognized Indians as wards of the state with rights similar to those of minor children. When the Portuguese invaded Brazil in the 1500s, there were five million Indians. Now there are approximately 200,000.[8]

When we finally entered Brazil from Venezuela and Colombia after our ten-day voyage, we began to see the ravages of civilization. The Brazilian government has appropriated large tracts of jungle to poor peasants from the cities. The forest has been destroyed, and the dead giants are left rotting on the ground as the peasants attempt to make a living by planting spindly banana plants and other trees among the forest debris. The trouble is that the soil of the Amazon is relatively sterile, and once the forest disappears, the humus composed of ever-recycling dead leaves and bacteria is eliminated. Because there is no ongoing regenerative composting of the soil, it becomes unsuitable for agriculture. The peasants are struggling to make a living, and the forest is dying.

The World Bank and the International Monetary Fund have, in their misguided beneficence, built a road called BR364 right through the middle of the jungle for "development" at a cost of $457 million. Side roads are then constructed, and the whole forest is opened up for destruction.[9]

The Amazon is threatened by several other major destructive enterprises.

LUMBERING Brazil owes a huge debt to U.S. banks. The World Bank and International Monetary Fund, which virtually represent the interests of these banks, have advised the Brazilian government to chop down the Amazon forest to pay back the debt. They have also funded Brazil's military dictators to build dams and encourage resettlement of thousands of people from the destroyed jungle.[10] Commercial logging is the second-most-serious cause of deforestation. Slash-and-burn destruction for agriculture ranks number one. In a single day in 1988, six thousand separate fires were burning in the Amazon forest—all man-made.[11] In 1980, it was estimated that the burning of tropical forests added an extra 1.7 billion tons of carbon dioxide to the air every year.[12] Logging is difficult in dense jungle, and often twenty trees are destroyed to obtain one suitable for timber, and one-third of the felled trees are destroyed to obtain access for logging equipment. The harvested timber is utilized for charcoal and firewood, for paper production, and for the building industry in First World countries.[13] Japan imports over 40 percent of the world's tropical timber.[14] It consumes the equivalent of one forest per day in disposable wooden chopsticks, or some twenty billion per year.[15] Much beautiful tropical wood is used in Japan as disposable plywood moldings for concrete in the construction business.[16] The United States imports two billion dollars worth of tropical wood per year.

MINING The soil beneath the Amazon forest houses valuable gold deposits, and half a million people are at present digging up and destroying the forest in their quest to find this treasure. Brazil is now the fifth-largest gold producer in the world; in 1988, over 182 tons of gold were extracted from the jungle floor. Mercury is used for the gold extraction process, and this toxic heavy metal is then released into the river. Indigenous populations living downstream rely on this polluted water for drinking and washing. Mercury poisoning is very serious; it causes kidney damage and Minimata disease, as well as brain damage and mental retardation in babies and children.[17] The forest also contains the richest iron ore deposits in the world. Once extracted, the iron is smelted in fires fueled by charcoal, made from partly burned trees. By 1992, seven smelting factories will be consuming 700,000 tons of charcoal per year.[18]

DAMS The World Bank recently planned the construction of 125 hydroelectric dams in the jungle by the year 2010, which would flood 600,000 acres of rain forest. Environmentalists have overturned some of these plans.[19]

CATTLE RANCHING This activity occupies 72 percent of the cleared forest areas, and most of the beef is used to supply fast-food burger chains in the United States, Central America, and Europe. Amazon cattle are cheap to raise, but the fertility of the cleared land is transitory. Good grass grows for only two years before the cleared area becomes a virtual desert, so more forest is sacrificed to grow more cheap beef for two more years.[20] Christopher Uhl, of Pennsylvania State University, estimates that a quarter-pound hamburger derived from steers raised in Central America represents the loss of fifty-five square feet of immeasurably valuable tropical rain forest containing one giant tree, about fifty smaller trees, twenty to thirty different tree species, about a hundred species of insects, and many other bird, mammalian, and reptilian species, as well as a huge diversity of fungi, lichens, mosses, and bacteria—in short, millions of living organisms and thousands of species are sacrificed for one quarter-pound hamburger.[21] Similar figures hold true for cattle raised on soil derived from the Amazon jungle.

DRUGS Cocaine accounts for the destruction of 1.7 million acres of Amazon forest in Peru, or about a tenth of the total deforestation in that country during this century. The demand for cocaine and crack in the United States is vast, and as long as it continues, there will be suppliers. Furthermore, the traditionally very poor peasants of Peru make a good living by growing cocaine.[22] (It should be noted here that, according to a number of sources, among them the Christic Institute in Washington, the CIA imports into the United States one-third to one-half of the hard drugs, in order to fund its ongoing covert operations.) In Bolivia, 38,000 tons of toxic waste, containing forty-one different chemicals, are dumped into tributaries of the Amazon River each year in the process of cocaine production. This could turn the Amazon basin into an ecological disaster area. The chemicals include 309 tons of sulfuric acid, 7,000 tons of calcium sulfate, and 3 million gallons of paraffin, which is almost not biodegradable and which asphyxiates fish and plankton.[23]

The global tropical rain forests are being destroyed at the rate of sixty acres, or sixty football fields, per minute. Every sixteen minutes, a rain forest equal in area to New York's Central Park is destroyed.[24] An area the size of Pennsylvania is cleared each year.[25] These forests contain or house about 50 to 80 percent of the world's species of plants and animals, of which there are estimated to be 30 million.[26] At the present, frantic rate of deforestation, all the world's tropical forests will within twenty-five to fifty years be destroyed, along with 15 to 24 million species, and the land will be desert. The Amazon jungle is thought to house 1,600 bird species, and just two and a half acres of jungle may contain 40,000 different species of insects. Rain forests are the oldest living ecological systems on earth, which have evolved over millions of years.[27] The Club of Earth maintains that species extinction is "a threat to civilization second only to the threat of nuclear war."[28]

One-quarter of the prescription drugs we use in medicine are derived from naturally occurring alkaloids and poisons in plants, and others are synthesized to resemble these natural chemicals. Some chemotherapeutic agents used in treating cancer are extracted from rain forest plants, as are antihypertensives, anesthetics, muscle relaxants, and contraceptives. Fourteen hundred rain forest plants contain anticancer chemicals.[29] We have really only begun to explore the natural riches of the rain forest.

Tropical fruits and nuts, oils, rubber and medicines, and natural pesticides and predators of crop-attacking insects offer an enormous and virtually untapped economic resource, potentially two to three times greater than all the timber of the forest. All these products are renewable and sustainable in perpetuity, requiring no fertilizers, pesticides, weedicides, fungicides, machinery, or fossil fuels to produce. Obviously, this should be seen as the economic wave of the future for the Brazilian government. Saving and nurturing the indigenous forest will be far more profitable than logging it; the trees will be left standing, and the lungs of North and South America will be saved.[30]

In the past, forests were systematically destroyed in times of war. The American military was responsible for a vivid example of deforestation during the Vietnam War. It sprayed nineteen million gallons of Agent Orange over huge areas of rain forest to expose the "enemy."[31] Agent Orange is contaminated with dioxin, one of the most carcinogenic agents known. When I visited Vietnam in 1986, huge mountains still looked like the backs of dinosaurs, blackened and scarred twelve years after the war's end. The only trees that would grow on these poisoned lands were Australian gum trees. Indigenous wildlife had been decimated along with the forest, and people were developing tumors in abnormally high numbers. The incidence of birth defects had also increased above normal levels, almost certainly induced by dioxin poisoning.

Tropical rain forests occupy 7 percent of the land surface of the globe.[32] All are under threat, in countries such as Indonesia, Borneo, Malaysia, the Philippines, New Guinea, India, Ghana, and Australia. In Malaysia, the survival of a small indigenous tribe called the Penans is under threat as international companies log twenty-four hours a day, at night under huge arc lights. Forests are being "mined," and for every ten acres destroyed, only one is replanted. Once the forest is gone, the Penan people may physically survive, but their culture, traditions, and religion will be shattered.[33]

During the rule of President Ferdinand Marcos in the Philippines, thousands of acres of tropical forest were sold and logged to provide money for Marcos and his friends and relatives. In New Guinea, Japanese companies bribe and coerce the government and indigenous people to allow clear-felling deforestation. In Haiti, which once boasted large areas of forest, only 2 percent of it remains. The Central American forests covered 60 percent of the land in 1960; clearing has reduced coverage to 33 percent today, while the First World has built lavatory seats, bookshelves, and houses from this beautiful unrenewable timber. The time to stop this waste is now. The planet cannot afford this reckless destruction.[34]

Forests are not just dying directly as a result of logging and burning; they are also threatened by indirect dangers, brought on our way of life.

ACID RAIN

When flying over the great expanses of Canada, I look down upon beautiful lakes that dot the landscape and realize that though many are crystal clear, they are biologically dead. The water's pH (a measure of its acidity) is so low that all living organisms have disappeared—fish, crustaceans, mollusks, and algae. Canadian lakes have been most severely polluted by acid rain originating in U.S. factories and power plants. Very high chimneys have been built in the Midwest to ensure that the pollution does not fall locally but, instead, travels thousands of miles in wind currents before depositing on distant lands.

Rain becomes acidic when sulfur dioxide and nitrous oxide gases, which are products of the burning of fossil fuels, combine with atmospheric water vapor to form sulfuric and nitric acids. These acids return to earth either as snow, rain, or fog or as particulate matter injected into the air by power plant chimneys.

Acid rain has devastated forests around the globe. In New England, the maple syrup crop has been halved in recent years because the magnificent maple trees are dying. Fifty percent of Germany's Black Forest has been killed, and similar figures pertain elsewhere—to the East Coast forests of North America, to other European forests, and to the forests in China and South America. Eighty percent of the lakes in Norway are dead because of acid rain. Sixty-seven percent of the trees in Britain are sick from its effects. Even the Arctic Circle is now covered with a brown haze of smog, and acid rain and snow—originating in European, Soviet, and Chinese factories—is polluting the lakes, rivers, and trees of Alaska. No country is immune from pollution, because acid rain does not respect national boundaries or artificial lines drawn on maps.[35]

As we have seen, although naturally formed ozone is necessary in the stratosphere to protect the biosphere from ultraviolet light, man-made tropospheric ozone is deleterious to plants. This lower ozone layer is formed when sunlight reacts with car and factory exhausts (smog). In addition to aggravating asthma and chronic respiratory disease in human beings, ground-level ozone damages crops and trees by inhibiting photosynthesis. Unfortunately, the harmful effects of ozone potentiate the plant damage caused by acid rain. So forests and trees suffer even more severely than would be expected, because the detrimental effect of one chemical enhances that of the other. The average levels of ozone in the cities of Europe and North America are three times above the minimum level that causes crop damage. Scientists believe that low-level tropospheric ozone has contributed to the death of 87 percent of the ponderosa and Jeffrey pines in southern California.[36]

Acid rain is also destroying our cultural heritage. Over the last twenty years, ancient statues in Athens, Rome, and other cities, which remained intact for thousands of years, are melting. As noses and ears dissolve in the acid rain, they become faceless. Many ancient buildings have eroded to become featureless, and acid rain is rapidly corroding metal structures such as the Statue of Liberty and railroad tracks in Poland. The Parthenon, in Athens, the wondrous Taj Mahal, in India, and the beautiful Cologne cathedral are all sustaining damage.[37]

We are damaging our heritage and the earth's ecology because we like to drive cars, live in affluence, and use enormous quantities of electricity. Obviously, the answer to this set of problems is to modify our life-style to a point where the notions of growth, profit, and affluence are obsolete.

Living in affluence does not necessarily make us happier than we would be if we lived close to the land, grew our own food, rode bicycles, and read by candlelight. After all, Dickens, Shakespeare, Beethoven, and Brahms wrote by candlelight. I am not suggesting we all need to do this, but some people may want to. Conservation and alternative energy are certainly in order. Television should be almost discarded because the endless advertising tends to encourage this destructive life-style; in any case, we learn very little from most TV programs. Better to indulge in family dynamics, to play cards and chess, to sit around the piano singing and playing, than curl up and numb our brains, our feelings, and our powers of critical thinking by watching the boob tube. However, if television were used in a responsible fashion by the corporations who own this medium, it could become a wonderful educational tool.

Dead trees have two main industrial uses—for paper and for building material. Let us first discuss the role of paper in today's society. When I was a little girl, the newspapers were small in bulk and packed with news. Now the newspapers are bulky, because they carry thousands of classified ads and full-page ads for cigarettes and the like. The news coverage is sparse, and many papers resemble TV journalism, for the photos are colored and the news bites are one or two paragraphs with no background information. The *New York Times* weighs several pounds on Sunday, full of advertising that very few people read. How many trees or forests are destroyed to produce one week's issues of the *Times,* let alone all the other newspapers in the country or the world! Very few newspapers are printed on recycled paper, and most are sent to garbage dumps—there to be burned and converted back to carbon dioxide or to rot gradually over many years. Recycling a single run of the *New York Times* would save 75,000 trees.[38] It is encouraging that New York City has recently instituted a recycling program.

Newsstands are really magazine supermarkets these days—packed with journals on diverse subjects ranging from pornography to knitting, automechanics, boating, racing, gardening, fashions, and cooking. Most of the glossy pages are covered with advertisements; only one-third to one-half of the paper contains articles, and most of these are somewhat superficial and trivial. Glossy paper for such publications is produced from short-fiber hardwood, including eucalyptus trees from Australian forests. Until recently, this timber was thought to have no economic value because it is extremely hard and cracks and splits, and because in many cases it is almost impossible to drive a nail into it. But over the last twenty years these forests have become very lucrative, and the Japanese have moved in to log Australia's last remaining wildlife habitats.

We must become conscious of our profligate use of paper. Consider the junk mail that fills our mailboxes each day: send it back to its source. You will find that the volume of this useless mail will soon decrease, and paper will be saved.

We blow our noses and wipe our tears on trees instead of on cotton. How many trees are felled to produce the quantity of paper tissues we use every year? Handkerchiefs, reusable for years, are hygienic and eminently sensible in these days of forest degradation. Instead of paper towels in the kitchen and paper serviettes at dinner, reusable cotton towels and table napkins are better choices.

Tampons are made from trees and bleached white by chlorine. When chlorine combines with paper pulp, it produces dioxin—one of the most mutagenic and carcinogenic agents known. It is therefore medically contraindicated for women to place a white paper tampon in their vagina, because the dioxin could well be absorbed through the vaginal mucosa to enter the bloodstream and settle in a specific tissue. Five to fifty years later, they may develop a cancer. The cancer will not wear an identification flag proclaiming its distant origin.

Furthermore, it is inappropriate to package tampons in plastic applicators that are discarded in the sewage system and that wash up on beaches and last for five hundred years. Tampons can be efficiently inserted by means of the fingers. In Sweden, unbleached tampons are used, but even these are made from trees. A useful alternative is a cosmetic sponge, which can be inserted dry, removed every few hours, thoroughly washed and squeezed dry, and then reinserted. Many women in Australia use this technique to soak up their menstrual flow.

Babies are swathed in white paper diapers covered with plastic. It is dangerous to place dioxin next to a baby's delicate skin, particularly if the skin has been treated with Vaseline or an oil-based preparation to help alleviate diaper rash. The oil enhances the absorption of dioxin. When I was twenty-seven, I had three babies under the age of three years, and they wore cotton diapers that I washed and hung on a line in the sun to dry. They smelled delicious, and I liked nurturing my babies by washing their diapers. From a public health perspective, it is significant that feces were disposed of down the toilet, whereas paper diapers full of human excreta are sent to the dump, where pathogenic bacteria could well contaminate drinking-water supplies.

Tea is served these days in white paper dioxin-containing bags made from trees. The tea bags fizz when placed in hot water—have you noticed? Now, tea leaves should be placed in a teapot and covered with boiling water; they never fizz. Often, in classy restaurants, the tea bags are packaged in aluminum foil bags, arranged in wooden boxes. Aluminum is manufactured from bauxite ore. Large quantities of electricity are used to smelt bauxite, thus adding to the greenhouse carbon dioxide. It is quite unnecessary to put tea in all this packaging.

Tea bags are but one example of grossly excessive packaging. When I was young, I used to ride my bicycle to the grocer's shop and ask for one pound of sugar. The nice grocer man ladled the sugar from a big jute bag into a brown paper bag, which he placed on a pair of scales. He weighed the sugar and closed the bag securely, and I rode home mission accomplished. No cars, no greenhouse gases, and no superfluous packaging.

A visit to the supermarket today offers a lesson in redundancy. Every item is packaged and covered with plastic, paper, cardboard, or foil. Packaging companies make large fortunes as we unthinkingly purchase their products. The next time you shop, take your baskets and string bags, and at the checkout counter unpackage the eggs, tomatoes, and sugar down to the paper bag and so on, place them in your baskets, and leave the packaging strewn over the counter. Encourage your friends, relatives, and neighbors to do the same, and soon the supermarket management will be bulk buying and discarding packaged goods to foster customer satisfaction.

Word processors eat paper and use it in a most inefficient way. Businesses and universities stress communication, but instead of talking to one another, participants cover reams of paper with redundant and unimportant memorandums. This wasteful use of technology must change.

All paper must be recycled. Newspapers and magazines must

cut down on advertising, they must be printed on recycled paper, and we must be discriminating about buying unnecessary and superficial magazines and papers.

Paper production is environmentally dangerous. Many toxic chemicals are made and used during the manufacturing process, and paper mills smell foul, of rotten eggs. Fishermen in Canada strongly advise Australians not to allow construction of paper mills, because the effluent destroys the surrounding countryside and kills the fish. I drove through the town of Lewiston, Maine, recently. There we crossed over a bridge that spanned a large river and upstream waterfall. This sight must once have been a tourist attraction, but the water splashing over the rocks was green-brown and smelled foul. It emanated from several paper mills upstream. I was surprised to learn that almost all the forests in Maine are owned by private paper companies and are therefore in peril. A local conservationist told me that this is true in most of the beautiful New England states, which derive a large part of their income from the tourists who are drawn to walk through the magnificent golden and crimson forests in October and November. I wonder how many New Englanders understand how precarious their natural beauty is.

Paper production has been shown to be medically dangerous to workers. An analysis of 1,071 deaths among paper mill and pulp workers in New Hampshire from 1975 to 1985 showed an increased incidence of cancer of the digestive tract and lymphoid system.[39] These workers were occupationally exposed to many dangerous organic chemicals, including dioxin, that are released into rivers and lakes, and that pollute the water and concentrate in fish. Interestingly, in August 1991 William Reilly, administrator of the EPA, issued a report indicating that dioxin is not as carcinogenic as was originally thought, and the EPA has opened a year-long evaluation of the risks associated with dioxin. However, Ellen K. Silbergeld, professor of pathology at the University of Maryland at Baltimore, strongly disagreed and said, "Nothing that has been learned about dioxin since 1985 when the EPA first published its risk assessment finding on dioxin in the environment supports a revision of science-based policy or action." Other environmental scientists suggest that the government is with the revisions attempting to protect not people but the industries that manufacture dioxin.[40]

It is possible to make paper from fibrous plants such as bamboo, hemp, papyrus, and kenaf. These crops can be grown on degraded farmland and pastures, and forests can thus be saved. Trees can also be grown as crops and harvested when appropriate. We need to plant billions of trees to save the atmosphere. Let's cover the earth with trees and bring back the birds!

Reliance on timber for building is appropriate in one sense, because the carbon is not released as carbon dioxide, but this wood must not be taken from virgin forests, only from timber plantations. We must explore other building materials, such as mud bricks, that are made from soil adjacent to the building site and dried by solar power. They are widely used in Australian and New Mexican adobe houses, with beautiful results.

I realize that you may be feeling quite depressed and somewhat helpless as you read this litany of tragedy. Let me therefore inspire you with some examples of human courage and heroism.

Tasmania is a small island state located off the southeast corner of Australia. It is heavily timbered and has a small population. The forestry department has for many years conducted logging operations in the forests, and the hydroelectric commission is fond of creating dams and flooding large areas of bush land. Some of the world's tallest flowering plants (trees) grow in this state.

In the early 1980s, it was decided to dam a large section of the Franklin River, which covers one of the most beautiful ecosystems in Tasmania. The people of Australia were incensed. Blockades and sit-ins were organized, and the area was occupied for months. The weather was often bone-chillingly cold, but people came from all corners of Australia to demonstrate. One gentleman, who had just retired as assistant secretary of defense, and his elegant English wife traveled from the mainland to protest, and they were duly arrested. The actions were efficiently coordinated, and the Forestry Commission workers never knew what would greet them as they moved their bulldozers and trucks into wilderness areas. Two-way radios, tents, wet-weather gear, food, and other equipment were donated by sympathetic people around the country. One of the outstanding heroes during this operation was a medical doctor named Bob Brown, who, dressed in a suit and tie, sat in a tree for several weeks and gave erudite and inspirational interviews to the press. Eventually, the people of Australia prevailed, a new national government was elected because of the protest, and the Franklin was saved.

However, the Tasmanian forests are still under threat. In 1989, two large paper companies (one Canadian) decided to build a large paper-manufacturing plant at Wesley Vale, a beautiful little fishing and farming village on the ocean. A local schoolteacher named Christine Milne, who had grown up in the area, decided that her village was not to be polluted by dioxin, organochlorines, or the smell of rotten eggs (hydrogen sulfide). She became the leader of a huge campaign that culminated in the cancellation of the plans to build the mill. Local conservative farmers, housewives, professionals, and children rallied around Christine, who grew so respected that she was subsequently elected as a green independent at the next state election—one of five who hold the balance of power in the Tasmanian parliament, where she can influence legislation to protect the forests.

Individuals can lead their countries and create movements to save the world. If you are sufficiently inspired to be one of these leaders, all you need is an extensive knowledge of the pertinent facts and a clear determination that you will prevail to save your community and, indirectly, your country and the world. Maintain your naïveté and determination, never procrastinate or become cynical or self-doubting, and you will win every time.

4

The Witches' Caldron: Toxic Pollution

When I was a child, women wore silk stockings and silk, rayon, or cotton underwear; floors were covered with linoleum or woolen carpets; china plates, metal cutlery, and glass were used in the kitchen; and garbage consisted of food scraps wrapped in old newspapers. The rubbish bin was a small lidded metal can that was put out weekly for collection, and once a week the iceman placed a block of lovely ice in the top of our ice chest, which kept the food cold. The milk was delivered at five each morning by a milkman driving a horse and cart. Lovely fresh bread was delivered at lunchtime, and Dad grew most of our fruit and vegetables. On the weekends, he walked the streets with a bucket and spade collecting horse manure to fertilize the garden. Our life was healthy and relatively free of artificial chemicals.

When I was eight, someone showed me a "nylon" doorknob. It looked like a magic material, and intuitively I knew it heralded the dawn of a new era—the plastic age. With some trepidation, I watched the introduction of Laminex bench tops, vinyl floor tiles, plastic lavatory seats, plastic-covered furniture, plastic piping for plumbing, plastic-filled planes and cars, disposable plastic kitchen utensils, and plastic medical equipment. The gravity of the situation became apparent when I was a medical resident in 1972. I asked a hospital administrator why we used disposable plastic scissors, scalpels, forceps, and so on, instead of metal surgical instruments. He told me that it was cheaper to buy plastic and throw it away than to employ people to wash steel implements.

Twenty years later, the situation is totally out of hand. I have witnessed the takeover of my world by plastic. In the hotel rooms I stay in on my travels, there is a plastic sachet full of nonbiodegradable shampoo, a plastic shower cap, and two plastic cups all wrapped in plastic.

Meals on airplanes illustrate the situation. Food is served on plastic plates covered with clear plastic wrap (a substance that emits a very toxic carcinogenic component called vinyl chloride, particularly when heated in a microwave oven). Tea is served in Styrofoam cups, which were foamed with CFC gas and which may contain untreated styrene, a carcinogen. These containers last for five hundred years. Milk is presented in a white cardboard carton contaminated with dioxin, and all the food tastes strangely of chemicals. Following the meal, the hostess walks down the aisle collecting all "disposable" utensils in a big plastic bag, which will be deposited in a garbage dump. About 25 to 30 percent of the volume of municipal garbage is plastic. This material is not biodegradable, and it remains in its original form for half a millennium.[1] Everything we use is eventually buried in the soil—plastic, cars, refrigerators, houses, tables, diapers, even our own bodies. I wonder how future archaeologists will classify our particular form of civilization when they analyze our garbage.

Most plastic is very difficult, if not impossible, to recycle. It is synthesized from oil, and scientists have crafted many different chemical molecules to make hundreds of forms of plastic endowed with physical properties like flexibility, strength, and transparency. When these different varieties of plastic are melted, the molecules change their physical characteristics, so it is virtually impossible to recycle plastics in their original form. Although some specific plastics are recyclable, most are not.[2]

Fast-food chains demonstrate the profligate use of plastic by modern societies. The main agenda of these chains is packaging, and their secondary agenda is food. The volume of packaging far exceeds the volume of food, and almost all this material is either plastic or cardboard derived from trees. Furthermore, the food they serve is not very nutritious, but it produces a sense of "satiation"—it is so laden with fat that the stomach empties slowly and one feels full and satisfied for longer periods of time than if one had eaten good high-protein and carbohydrate food. Fat is cheap, so it is economically advantageous to sell potatoes (carbohydrate) and hamburgers (protein) soaked in fat, although McDonald's introduced a low-fat hamburger in the spring of 1991. This same corporation generates three hundred tons of trash daily at its 8,500 restaurants in the United States—most of it paper and cardboard packaging made from trees. There are thousands more McDonald's in other parts of the world.[3]

During the manufacture of plastic, large quantities of different chemicals are made as by-products. They constitute toxic waste that is emitted into the air through factory chimneys, poured into sewage systems, sent to garbage dumps, drained into streams, rivers, and lakes, buried in landfills, or illegally dumped by the Mafia at night, when no one is looking. The food chain concentrates many of these toxic organic chemicals, so the plastic we use every day may come back to haunt us as poisonous food, water, or air, or it stays in our garbage dumps for hundreds of years.

There are other interesting ways to dispose of toxic waste; the following story describes one of the more innovative. An attorney, Mark Pollack, ten years ago attended one of my speeches in which I addressed the medical effects of nuclear war. He left the lecture saying to himself, "Well, I'm either part of the problem or part of the solution." He decided to create a new legal specialty and opened a practice in which he concentrated on toxic-waste issues. He sues firms for willful and knowing contamination of the environment. For example, he sued Shell Oil for dumping toxic wastes into San Francisco Bay and won $18 million.

Early in 1990, he was called by a woman in Napa Valley who

was worried that a lake near her house was blue. He told her not to be ridiculous—that lakes were not blue—but as she was insistent, he decided to investigate. The lake was indeed blue. He noticed that adjacent to the lake stood a large building. The man who opened the door when he knocked had blue material running from his nostrils. He said that he manufactured a blue chemical, some of which he drained into the lake. As Mark stood in the office, he noticed two lots of large barrels stacked in the corners of the room. He asked what the barrels contained, and the man informed him that he had been paid to store them by chemical companies who wished to dispose of their toxic waste. On further questioning, he became quite enthusiastic and said he mixed the two chemicals together to make a thick paste. He then sold this paste to certain mortuaries, where it was stuffed into the thoracic and abdominal cavities of cadavers after autopsy, when the organs have been removed. Thereupon the bodies were dressed in their best clothes and buried after appropriate funeral rites. These bodies then became repositories of toxic waste, and the disposal problem of the chemical companies was solved. This anecdote may give a clue to many other illegal means of waste storage.

In my experience, if you imagine the very worst possibility related to white-collar corporate crime, it has probably been done somewhere by someone at some time. It is important to understand that many of us live in our toxic waste every day.

A young woman in Los Angeles said to me the other day, "It's a good day today; you can see through the air!" The people of Los Angeles remind me of the frog experiment. If you put a frog in cold water and heat the water gradually, the frog does not notice the temperature change and eventually boils to death. If you drop a frog into boiling water, it leaps out and saves its life.

A recent study performed in Los Angeles at the University of Southern California showed that in the lungs of one hundred youths who had died violent accidental but nonmedical deaths, 80 percent had "notable abnormalities in lung tissue, and 27% had severe pathology." The pathologist Russell P. Sherwin, who performed the study, said, "The youths were running out of lung. . . . Even if I were to assume that most of these people were smokers, I'm seeing much more damage than I would expect to see."[4]

About 125 million pounds of toxins are discharged into the Los Angeles air each year, including the potent carcinogens trichlorethene, methylene chloride, and benzene. Some 46 to 87 percent of this material comes from the exhausts of cars, buses, and trucks, and 62 percent is produced by 25 percent of the biggest corporate polluters.[5]

On Earth Day 1990 (April 22), I noted a two-page advertisement in the *New York Times* proclaiming the value of Earth Day. On one page was a beautiful photograph of our fragile planet taken from space, with a caption cautioning responsible management of the earth. Filling the opposite page was a list of 170 major U.S. chemical companies that had commissioned the advertisement, including Dow Chemical, Monsanto, and Ciba-Geigy. These same companies make chemicals for commercial use and produce the toxic waste that poisons our air, water, soil, and food. They have clearly become conscious of people's fear about toxic chemicals, but they obviously seek to capitalize on our anxiety and to convince us that they now make "environmentally friendly carcinogens."

Earth Day was very successful and attracted hundreds of thousands of concerned citizens, but the community activities were, to a large degree, underwritten by the very corporations that are causing the problem. It is important that we think critically about who pays for what. Cigarette companies sponsor sports events and opera, chemical companies sponsor Earth Day, and nuclear power companies sponsor local community organizations and kindergartens. Meanwhile, President Bush has been asking corporations to underwrite school programs.

Most corporations seem to be hopping onto the green-movement bandwagon, but caution is indicated because they do not necessarily practice honesty. For instance, British Petroleum now paints its gas pumps and oil tankers bright green, so if a tanker accidentally spills millions of gallons of crude oil into an environmentally sensitive bay, everything should be okay because the ship was painted green. This comment is facetious but basically true.

I am also skeptical about "environmentally friendly" products that appear on supermarket shelves. The list of chemicals often contains toxins. Even if the product seems to be chemically innocuous, large amounts of electricity may have been used in its production. The ecological claims made on the labels tend to be fraudulent—degradable plastic, recycled plastic, "decomposed" diapers, and so on.

To date, seven million artificial chemical compounds have been synthesized. Most have been noted only once in the chemical literature. Many are released deliberately into the environment precisely because they are toxic—to kill weeds, trees, and insects. More than $13 billion worth of pesticides are used globally.[6] About 80,000 are now in common use, almost all of them toxic. We continue to add 1,000 to 2,000 new chemicals each year to this frightening inventory, and very few are methodically tested for carcinogenicity or mutagenicity.[7] No information is available on the human toxicity of 63,200 of these commonly used chemicals, and a complete toxicology profile has been prepared for only 1,600.[8] All of them derive from scientific research and development. Many are necessary components of industrial production, and others are unwanted by-products.

The number and quantity of chemicals in common use boggle the imagination. Industry produced seven times more goods in 1990 than in 1960, and global production of organic chemicals increased from 1 million tons per year in the 1930s to 250 million tons in 1985. Annual production is now doubling every seven to eight years.[9]

Because of our own addictive consumption, people in the First World cause 100 to 1,000 times more pollution per capita than people in the developing world. (The U.S. population, a mere 4 or 5 percent of the world total, creates half of the world's toxic waste.)[10] Of the 375 to 500 million metric tons of hazardous waste manufactured on the planet each year, the United States produces 260 million metric tons, or more than one ton per person.[11] This pollution is a natural consequence of advertising, lack of environmental education on the part of the scientific community, and an extremely high standard of living, which contributes, over time, to much sickness and suffering.

Dangerous chemicals are found in almost every household article. Carcinogens are used to dry-clean clothes, and traces of chemicals often remain on the garments when we collect them from the shop—you can actually smell them. Paints and thinners are supplemented with toxic chemicals designed to discourage fungal growth or act as quick-drying agents. Cancer-causing chemicals are used to kill termites, cockroaches, and insects in homes, offices, schools, hospitals, and restaurants. Cleaners and deodorizers contain toxic chemicals, as do garden pesticides and fungicides. Toxic fumes emanate from synthetic carpets, furnishings, and curtains. Carcinogenic formaldehyde leaches into the air from certain types of insulation and fabrics used in walls and ceilings.

I was unsuspectingly poisoned recently when I painted my chimney with lacquer paint. I used turpentine to wash the paint off the brush and my skin, and the next day I awoke with a splitting headache. By midafternoon, my skin was covered with a raised nodular rash that was not itchy. I was perplexed about the diagnosis and thought I was infected with a strange virus until I realized that I had been poisoned by the paint. The turpentine had obviously facilitated absorption of the toxic chemical or chemicals. There had been no warning on the label, and I was horrified to learn that exotic chemicals are added to ordinary household paint. The symptoms took several days to subside.

Physicians are surprisingly uninformed about the medical complications of most toxic products in household cleaners, sprays, paints, and detergents. This subject is not an integral part of the medical school curriculum, yet most bottles of household chemical carry the warning "If swallowed or inhaled, contact your physician."

I am a physician who is quite aware of the dangers of toxins, but when I was invited by the Australian Cotton Growers' Association to talk at its annual meeting about the medical dangers of chemicals used in its industry, I realized I was totally ignorant. I embarked on a three-week intensive study program that became an exercise in frustration. I collected literature from the Department of Agriculture, from the cotton industry, and from the Health Department. When I read the extensive list of organochlorines, organophosphates, dioxin-containing sprays, and other chemicals used as weedicides, fungicides, herbicides, pesticides, and fertilizers, I was shocked. I knew that many were dangerous, but I found it very difficult to obtain accurate and complete information on the path of these materials in the food chain and on the pharmacological consequences of the chemicals in the human body. It was immediately obvious from an ecological perspective that the soil and adjacent waterways would be poisoned by the sprays and that bees and birds would be damaged, but I was dismayed to discover that cottonseed oil is the main ingredient of cooking oil and margarine. When I called the State Health Department for data on chemical residues in the oil, I was informed that the appropriate tests had not been done. Then someone told me that women with tiny babies are advised to wash new cotton garments three to four times to remove chemicals adhering to the cotton fibers before they use the clothes.

I told the cotton growers and their wives that their industry appeared to be medically and ecologically dangerous, but that specific information about the chemicals was not readily available to physicians. I told them that if I, an environmentally conscious doctor, did not know much, my colleagues would know even less. They received this message unwillingly, because although they knew it to be fundamentally true, they made a good living from growing cotton.

Now let's talk about what happens to these toxic chemicals after we have finished using them. Their disposal poses an enormous problem. Garbage was relatively benign in the old days, because the use of chemicals was not widespread, but it is now dangerous stuff. We throw household cleaners, insecticides, spray cans, mothballs, paint thinners, bleaches, ballpoint pens, floor cleaners, plastics, detergents, oil, dry-cleaning chemicals, window cleaners, dioxin contained in white cardboard and paper, batteries containing lead and acid, cars containing heavy metals, plastics, and oil, and refrigerators containing CFC gas, to name a few, into the garbage.[12] The U.S. population each year discards 16 billion diapers, 1.6 billion pens, 2 billion razor blades, and 220 million car tires, along with enough aluminum to rebuild the U.S. commercial airline fleet four times over.[13]

The obvious solution to our dilemma is to use cloth diapers, nondisposable razors, recyclable glass bottles for milk and drinks (bottles that are not melted down and remade but that are washed and used time and again), bicarbonate of soda and vinegar for cleaning, ordinary soap instead of detergent, fountain pens rather than throw-away ballpoint pens, and biologically safe pesticides. We must build mass transit systems and, if necessary, discard small bicycle tires instead of car tires. In other words, we need not stay on this treadmill of addiction to hazardous consumption.

Municipal garbage dumps leach toxins into underground aquifers and nearby rivers and streams, thereby endangering wildlife and human food chains. So dangerous was municipal waste in 1990, that more than half the hazardous-waste dumps flagged by the congressionally appointed Superfund were municipal garbage dumps and landfills. The other hazardous sites were filled with toxic chemical waste from industry.

To quote from *The Global Ecology Handbook,* "8 out of 10 Americans live near a hazardous waste site." There are 15,000 uncontrolled hazardous-waste landfills and 80,000 contaminated lagoons in the States.[14]

One of the most famous toxic disasters was Love Canal, in New York State. In 1945, a Hooker Chemical Company analyst wrote in an internal report that the "quagmire at Love Canal will be a potential source of law suits," and he called it "a potential future hazard." Hooker dumped 40,000 metric tons of toxic chemicals, including dioxin, lindane, and arsenic trichloride, into Love Canal, which emptied into the Niagara River, adjacent to one of the world's greatest natural wonders—Niagara Falls. The company later filled in the canal and donated the site for the construction of an elementary school.[15]

President Jimmy Carter was forced to declare a state of emergency when toxic chemicals began oozing into basements of houses and when doctors warned that people could develop cancer, leukemia, and birth defects from these materials.[16] The company has recently been sued for $250 million, the estimated cleanup price. But *cleanup* is a euphemism. How can the toxic waste that entered the Niagara River over the past thirty years be retrieved? How will the people who died, are dying, or will die of chemically induced cancer be compensated? How much do you pay a woman who gives birth to a deformed baby, or the deformed person for his or her lifetime of disability?

Love Canal is but one of thousands of tragedies about to become manifest all over the country. By 1988, some 1,177 hazardous sites had been officially placed on the Environmental Protection Agency's priority list, which means that they are eligible for "cleanup" under the Superfund legislation, using federally allocated funds. Every state has at least one of these sites, but New Jersey has 110 and Pennsylvania 97, and New York, Michigan, and California have more than 75 sites each.[17]

Unfortunately, the situation is more grim than the official EPA figures would have us believe. In fact, the U.S. government is the nation's chief polluter; federal facilities discharge almost 2.5 million tons of toxic and radioactive waste without having to report a drop. According to the General Accounting Office, 95 percent, or 200 million tons, of all chemical pollution is still unreported because the federal government is exempt from reporting, the EPA is too weak in law enforcement, and the law has too many loopholes.[18] There are also 14,401 potentially contaminated toxic dump sites at Department of Defense facilities and the weapons manufacturing plants scattered around the United States.[19]

Despite this grave situation, industrial pollution is increasing. In 1987, some 10.5 billion pounds of toxic chemicals were released into the air, water, and soil of the United States, and more than half of the chemicals dumped into the country's waterways are not covered by EPA regulations.[20]

Huge fuel storage tanks buried at gas stations are also sources of a disturbing amount of underground pollution. Constructed of steel, they rust over time. Of the 1.4 million underground gas tanks, 15 percent are leaking. It takes just two pints of gasoline or oil to contaminate several million gallons of drinking water.[21] According to some estimates, one-third of the underground aquifer of the United States has been polluted with carcinogenic benzene or chemicals from these and other sources.

Airborne toxins are also a frightening consequence of our modern industrial society. Several vivid examples come to mind. St. Gabriel, a town on the Mississippi River, is host to twenty-six petrochemical factories, which belch 400 million pounds of chemicals, including the carcinogens benzene, carbon tetrachloride, chloroform, toluene, and ethylene oxide, into the air each year.[22]

In addition, half a million pounds of vinyl chloride gas were released in the St. Gabriel region in 1987.[23] Vinyl chloride is an extremely potent liver and brain carcinogen. It is used in the production of polyvinyl chloride, or PVC, the clear plastic used for bottles containing cooking oil, for plastic piping in plumbing, and for many other purposes. Plastic wrap is also made from PVC. PVC is not stable, and vinyl chloride gas tends to leach out of the plastic bottles and plastic wrap into the food, particularly in hot climates or warm houses and supermarkets. Vinyl chloride is fat soluble and is therefore likely to collect in fatty goods such as cooking oil, meat, and cheese that are covered with Saran wrap.

The East Coast between New Jersey and Boston is called "the toxic corridor." When I drive past the eerie petrochemical plants in New Jersey's "cancer alley," I am reminded of some surreal landscape out of science fiction. The air smells foul, there is no sign of vegetation or life, the waterways look like oil, little wooden houses are scattered among huge masses of pipes and chimneys belching fire, and yet real people obviously live and work amid and in these frightful monsters.

Another example of global airborne pollution is the two million tons of lead released into the air each year. This heavy metal concentrates in kidneys, causing renal damage, and destroys neurones, inducing brain damage and lowered IQ in children and adults.[24]

Pollution is not unique to the United States. In 1989, I stayed in a château in Ardèche, overlooking the Rhone Valley and the French Alps. Although the weather was clear and sunny, only rarely did I see the Alps, because the valley was so filled with smog that the pinnacles seldom pierced the pollution. On a recent train trip in northern Italy, I turned to the man beside me and said, "Look at that smog," and he replied, "That's not smog; that's fog." The air is so thickly polluted that it would be impossible for Renaissance artists to paint their pictures these days. The glorious landscapes and mountains have disappeared. Beautiful, tidy Germany is the same—cows graze in green paddocks amid a gray haze. Europe is disappearing from sight, drowning in its own pollution, the most visible sign of its wealth and good economic performance.

But if Western Europe is bad, Eastern Europe is impossible. Twenty years ago, public health specialists recognized that their population was endangered by industrial pollution, but Communist party officials quashed concern and public criticism. In some areas in Poland, the air was so polluted that people had to live underground in salt mines in order to breathe. In the industrial region of Katowice and Kraków, placentas showed high concentrations of lead, cadmium, mercury, and other heavy metals, and these placentas were taken only from healthy babies.[25]

In most Eastern bloc countries and parts of the Soviet Union the air quality has deteriorated sharply over the last two decades. Very few emission controls were ever used on factory chimneys, and those that were tended to be antiquated.[26]

Authorities deliberately did not perform surveys on the affected populations to determine whether detrimental effects were occurring. In Silesia, Poland, the growing of vegetables has now been banned in contaminated soil. The Czechoslovak Academy of Sciences declared that 50 percent of the country's drinking water was unacceptably polluted. In northern Bohemia, home to power plants and chemical factories, infant mortality is 12 percent above the norm. When Nikolai Ivanov, a physician, complained that a nearby copper plant was causing arsenic and lead poisoning in his patients in 1989, he was publicly criticized by a local Communist leader. In the industrial cities of Leipzig, Halle, and Dresden, in what was formerly East Germany, death rates from cancer and heart and lung disease are 15 to 25 percent higher than in Berlin.[27]

SOLUTIONS

Solutions do not abound. Dumping in soil or water is not the answer. Incineration is becoming a popular approach to the problem of disposal. If the organic chemicals are composed only of carbon and hydrogen, they are converted into carbon dioxide and water when incinerated. However, most toxic compounds emanating from plastics and the like contain chlorine, and some contain heavy metals. When subjected to high temperatures, the chlorine compounds are converted into dioxins, which escape from the chimney together with the heavy metals. Incineration is thus a flawed solution.

Mark Pollack continues to sue more corporations, and his actions should inspire other attorneys. Instead of suing doctors and each other, they can sue the people who really count—the toxic polluters of our planet.

Portland, Oregon, in 1990 passed a law to ban Styrofoam. The city employed an official, nicknamed the "Styro-cop," to enforce the law. He inspects fast-food outlets, restaurants, nightclubs, and supermarkets, levying fines of $250 for the first offense and $500 for a repeat violation. He is passionate about his job and says, "To use plastic to drink eight ounces of coffee for two minutes and throw it away where it will take up space forever is absurd."[28]

Basically, though, the only solution to the problem of poisonous chemical waste is to stop making the stuff. Recycling is not the solution, because it gives chemical corporations the green light to continue making more plastics, more aluminum, and so forth. Glass bottles must be returned in exchange for a deposit, washed, and reused. Glass must not be melted down and fashioned into new bottles, because this process produces more global-warming carbon dioxide gas. We must stop buying any food or drink that comes in disposable containers, whether made of glass, plastic, or aluminum. All institutions must use china, glass, and stainless steel kitchen and eating utensils. This ecologically sound policy will also provide employment for a greater number of people. Paper, on the other hand, can and must be recycled. We will have to learn to manage without plastic, quick-drying paints, artificial fibers, cars, CFC-run refrigerators, and the inventory of household chemicals and revert to the materials used by our grandmothers. The chores of daily life will be somewhat more time-consuming, but maybe we will find satisfaction and even great joy in a bit of hard work and a sense that we are nurturing the planet.

FOOD

We must be very conscious of the chemical composition of the food we eat. Our grandparents ate fruit that was not perfectly formed and perhaps contained a few worms, but it tasted delicious. Then farmers began using sprays to kill pests and increase their crop yield. Before long, chemical companies got into the act and agricultural products became a large portion of their business. Initially, the yields of chemically treated crops increased quite dramatically, but over time the predatory insects and parasites mutated and developed immunity to the sprays. The chemical companies therefore developed new and more exotic poisons to combat this human-induced immunity. Despite these carefully applied chemicals, however, crop losses caused by pests have actually increased since the 1940s, from 32 percent to 37 percent.[29] Meanwhile, the soil has been contaminated. As a result, worms and naturally occurring microorganisms that form the base of the pyramidal food chain have been killed and bees and birds that fertilize the crops have been poisoned.

Amazingly, less than 0.1 percent of all pesticides ever reach the target insects. The other 99.9 percent are dispersed in groundwater, lakes, weirs, soil, and the air. Groundwater has been contaminated with more than fifty varieties of pesticide in some thirty U.S. states.[30] The use of pesticides increased from 300 million pounds annually in 1966 to more than 1 billion pounds in 1981. And pesticide use globally is increasing by more than 12 percent. Somewhere between 400,000 and 2 million people on earth are poisoned by pesticides each year and 10,000 to 20,000 of these die. The National Academy of Sciences reports that pesticides in food may cause one million cases of cancer in this generation of Americans.[31] All this ecological damage to grow unblemished, artificially beautiful food! I believe it is better to eat a sweet, chemically uncontaminated apple containing a worm than a perfect toxic apple tasting like cardboard.

Farming must be made safe. Organic farming relying on mulch, compost, manure, and naturally occurring pesticides has been proven to be more productive than chemical farming. In Australia, organic produce is attracting prices 20 to 100 percent higher than those of chemically grown produce.[32]

The environment, then, is poisoned by industry, by waste dumps, and by agriculture. But we still must examine the transportation of toxic waste, accidents involving chemicals, sewage, and radioactive waste. I am aware that this chapter contains a litany of horrifying facts and statistics, but before we can cure a patient, we must make an accurate diagnosis. Before we can clean up our lives and cure the planet, we must recognize the extraordinary damage that our life-style inflicts on the planet.

ACCIDENTS

Accidents that disperse chemicals into the environment are now weekly items in world news. Trucks crash and burn, trains roll over and explode, oil and toxic chemicals leak from sinking ships, and fires and explosions engulf huge chemical factories. The major accidents that have occurred since 1959 are listed in the following table.

THE DIARY OF DISASTER

Year	Place	Cause	Casualties
1959	Minamata/Niigata, Japan	mercury discharged into waterways	400 dead, 2,000 injured
1973	Fort Wayne, U.S.	vinyl chloride involved in rail accident	4,500 evacuated
1974	Flixborough, U.K.	cyclohexane involved in explosion	23 dead, 104 injured, 3,000 evacuated
1976	Seveso, Italy	dioxin leakage	193 injured, 730 evacuated
1978	Los Alfaques, Spain	propylene spilled in transport accident	216 dead, 200 injured
	Xilatopec, Mexico	gas exploded in road accident	100 dead, 150 injured
	Manfredonia, Italy	ammonia released from plant	10,000 evacuated
1979	Three Mile Island, U.S.	nuclear reactor accident	200,000 evacuated
	Novosibirsk, USSR	accident at chemical plant	300 dead
	Mississauga, Canada	chlorine, butane released in rail accident	200,000 evacuated
1980	Sommerville, U.S.	phosphorus trichloride in rail accident	300 injured, many evacuated
	Barking, U.S.	sodium cyanide in plant fire	12 injured, 3,500 evacuated
1981	Tacoa, Venezuela	oil explosion	145 dead, 1,000 evacuated
1982	Taft, U.S.	acrolein involved in explosion	17,000 evacuated
1984	São Paulo, Brazil	pipeline explosion, petrol released	508 dead
	St J. Ixhuatepec, Mexico	gas released in tank explosion	452 dead, 4,248 injured, 300,000 evacuated
	Bhopal, India	leakage at pesticide plant	>2,500 dead, thousands injured, 200,000 evacuated

(Continued)

1986	Chernobyl, USSR	nuclear reactor accident	>25 dead, at least 300 injured, 90,000 evacuated and fallout spread over much of Europe
	Basel, Switzerland	fire at pesticide plant	Rhine seriously polluted
1987	Kotka, Finland	monochlorobenzene spilled in harbor	sea floor polluted

Source: United Nations Environment Programme Brief No. 4

Three recent disasters illustrate this problem. In 1986, the Chernobyl meltdown contaminated most countries in Europe (east and west). Radioactive elements that remain poisonous for hundreds to thousands of years rained down on some of the best agricultural land on that continent. We therefore must remember that food grown in these areas may well be radioactive for extended periods and that people eating this food will be at high risk for developing cancer, leukemia, or genetic disease.

In the same year, a dreadful spill of toxic chemicals into the Rhine River followed a fire in a chemical factory located in Basel, Switzerland. Thirty tons of pesticides so polluted the river that 500,000 fish and 150,000 eels were killed.[33]

In 1984, Union Carbide accidentally released huge amounts of methyl isocyanate gas from its plant in Bhopal, India, killing 3,300 people and injuring 20,000 others.[34] It later released the same chemical from its plant in West Virginia.

As environmental awareness becomes paramount in educated societies, chemical companies that wish to continue their business transfer their most polluting industries and toxic wastes to developing countries. Even Australia is the recipient of some of Japan's worst polluters. Third world countries are under enormous pressure to accept the unwanted toxic waste of the First World. In some countries, this provides a ready source of income. Sierra Leone was offered $25 million by a U.S. firm to store toxic waste. In 1980, West Germany solicited some North African countries to accept their hazardous waste, and a U.S.

corporation attempted to construct a hazardous-waste incinerator on a Caribbean island. Falsely labeled containers of poisonous chemicals are routinely shipped into developing countries, and some of them are being encouraged to construct their own waste-disposal facilities, which then become the toxic-waste sewage systems of the First World.[35] Lonely ships laden with toxic waste circle the globe for months searching for countries that will accept their cargo, only eventually to dump it surreptitiously at sea. These examples are far from unique.[36]

ACCIDENTS WITH OIL

Oil is pumped out of the ground, and millions of gallons are burned every day in our vehicles and industrial engines. This burning, as we have seen, produces global warming. As we transport the oil to and from refining plants, we spill it from ships and trucks and dump toxic by-products into pits and waste facilities that leach into the underground water. Out of oil, we make plastics and poisons that add to the degradation of the planet. It might seem better to leave oil in the ground, but if we really have to use it, it should be strictly rationed and the price of gasoline should reflect the damage it is wreaking upon our planet and our children's future.

The *Exxon Valdez* spill is a perfect recent example of the oil plague. On March 24, 1989, the *Exxon Valdez* oil tanker foundered on a reef in Prince William Sound, Alaska, and spilled eleven million gallons of sticky crude oil into a pristine wilderness waterway full of otters, salmon, herring, seabirds, and myriad other wildlife. It was many days before the Exxon Corporation organized itself to begin to deal with the mess, and the local fishermen watched with horror as otters scratched their eyes out and drowned in the oil, seabirds gagged on the toxic sludge, and birds staggered around the beach, their feathers caked in oil. People tried to wash the birds in water and detergent, but most died anyway.[37]

Fifty-eight scientific studies released by the National Oceanic and Atmospheric Administration in April 1991 revealed that the long-term ecological damage was far worse than had originally been estimated, that hundreds of thousands of birds and thousands of otters were killed, and that baby fish and fingerlings exhibited abnormally high rates of birth defects (demonstrating the mutagenic and teratogenic fetus-damaging effects of the toxic oil products—effects that will probably continue for many generations). Economists who participated in the study estimated the cost of damage to Prince William Sound to be $3 billion to $5 billion, yet Exxon was at first fined only $1 billion. A federal district judge, Russell Holland, has since ruled that the fine was inadequate.[38]

Coincidentally, at the same time that the fine was overruled, Exxon announced unprecedented profits for the first quarter of 1991. They were the highest made since John D. Rockefeller founded the company in 1882. Net income rose 75 percent, to $2.24 billion, largely because of the Persian Gulf war and sales of jet fuel, kerosene, and other refinery products to the military.[39]

But this is not the only ecological violation by Exxon and Texaco, its subsidiary. In 1989, it was fined $750,000 for failing to perform, and even fabricating, tests of oil-rig-blowout protection equipment. In the same year, it violated the Migratory Bird Treaty Act, which protects certain species of birds. In that year, more birds died in oil company waste pits than were killed by the *Exxon Valdez* spill. In 1988, a Texaco subsidiary was fined $8.95 million for failing to adequately clean up areas around five thousand corroded, leaking drums of toxic waste.[40]

As I have said, oil (and its products) is one of the major contaminants of the planet, yet the global community refuses to acknowledge this fact. So addicted are the industrial nations to a steady supply of oil that when Saddam Hussein threatened to control approximately 20 percent of the world's oil supply by invading Kuwait, the United States and its allies decided to force him out of Kuwait by military action. It is true that the United States had the support of the United Nations and some token military assistance from its allies, but the whole operation was masterminded and organized by George Bush and the Joint

Chiefs of Staff. Bush stated at the start of the Iraqi invasion of Kuwait that America was in the Persian Gulf to protect its "way of life." It is also important to note that this war came at a time when the Berlin Wall had collapsed, the cold war was over, and the stocks in the military-industrial corporations were seriously declining.

Meanwhile, a recent report from the Organization for Economic Cooperation and Development (OECD), which represents the world's richest nations, said that the industrialized nations, which make up 16 percent of the globe's population and inhabit 24 percent of its area, produce 72 percent of the world's gross product, own 78 percent of the cars, consume 50 percent of the energy, and import 73 percent of the forest products.[41]

These people, says the report, are intent on worsening pollution of the sea, the air, and the land by ever-increasing "development," which encourages the use of the private car and which leads to increased congestion and wider dispersion of leisure activities. The amount of waste each person creates by this lifestyle continues to increase; it now stands at ten tons per capita per year. Over the past twenty years, this population has become richer and it lives longer. This population includes us. Our selfish and greedy attitudes necessitated the war in the Persian Gulf.

I will not discuss the political ramifications of the Gulf war catastrophe, except to note that hundreds of thousands of people were killed by the most sophisticated weaponry in the history of the human race. The number of Iraqi soldiers killed has been estimated at 150,000; civilian deaths, at anywhere from 5,000 to 100,000. The actual numbers, including the number of children, may never be known.[42] The United States used more firepower in five weeks than was used either in World War II or in the Vietnam War. The air force flew more than 90,000 sorties—3,000 per day, or 1 per minute. At least 70 percent missed their targets. A UN mission has since reported that Iraq was bombed back to the "preindustrial age" and that the resultant situation was "near apocalyptic," with damaged sewage systems and water supplies, few hospital facilities or drugs, and little food. In some ways, to use Pentagon jargon, it was a "near-nuclear war."

In truth, it was not actually a war, because there was no retaliation; it was a massacre.

This war produced possibly the most devastating ecological consequences of any war in history. Immediately after the war, 732 oil wells were burning in Kuwait, set alight or damaged by departing Iraqi soldiers and America's weapons that missed their targets. Red Adair and other oil well fire specialists have been quelling the fires, and estimates of the time for completion of the task vary from two to ten years. By mid-September 1991, some 429 of the smaller wells had been recapped.[43] As the fires rage, it is almost dark at noon in Kuwait City, and much of the land in Kuwait and adjacent Iraq, Iran, and Saudi Arabia is covered with a greasy oil slick. Blackened snow has been found on the Himalayas and black soot at Mauna Loa, in Hawaii. It is possible that the fires will increase global carbon dioxide emissions by 10 percent. Sulfur and nitrogen oxides will cause far-ranging acid rain. Indeed, acid rain is "very severe" in Iran, Pakistan, Afghanistan, and the southern part of the Soviet Union—which are often downwind from Kuwait. According to the World Meteorological Organization (WMO), the fires each day discharge over 40,000 tons of sulfur dioxide into the atmosphere, equal to the combined emissions of Germany, France, and Britain.[44] However, some scientists say that the carbon dioxide emissions produced in the four months since the war ended were equivalent to 1 percent of the normal global production.

Burning oil also produces cyanide, dioxins, furans, PCBs, and other extremely poisonous, cancer-causing toxins and gases, which will affect populations living in the vicinity of the fires.[45] These poisons tend to cling to very fine particles, less than ten microns in diameter, which lodge in the tiny terminal airways in the lung after inspiration. The particles can stay in the lung for years, thus exposing surrounding cells to the cancer-causing chemicals. This is therefore potentially a very serious medical problem. In fact, the WMO estimates that one year of burning oil wells could double the quantity of these fine particulates in the world's atmosphere.

The fragile desert ecosystem of the Middle East was damaged

by the war, as was the extensive irrigation system of dams, dikes, and channels in the fertile Tigris and Euphrates river systems. Baghdad was one of the most ancient and beautiful cities on earth, and Iraq itself has attracted archaeologists because it is full of valuable ruins from ancient civilizations. Baghdad was bombed daily, some say even carpet bombed. Antiquities were destroyed in the name of power, oil, and control.[46]

A quantity of oil twenty-four to thirty-two times that of the *Exxon Valdez* spill was emptied into the Persian Gulf waters by Hussein and probably by U.S. bombs damaging oil facilities. This oil affected important desalination plants that line the gulf and supply eighteen million people with fresh drinking water. Three of them had to be closed down as the oil slick spread.[47]

Apart from the threat to the desalination plants, the most important side effect from the oil spill was the projected death of hundreds of thousands of seabirds and fish, of which the Persian Gulf has a rich variety. The feathers of seabirds have been gummed up by the oil, and toxic hydrocarbon chemicals dissolving into the water from the slick could threaten turtles, dolphins, whales, sea cows, and other fish and birds.[48]

Sea cows are called dugongs and, although they have tusks like those of elephants and look quite ugly, are such graceful, timid, inquisitive, and lovable animals that they inspired the original mermaid legends many years ago. These mammals, whose numbers in the Persian Gulf have been estimated at seven thousand, graze on beds of sea grass. They already lived in an environment vulnerable because of high salinity and high water temperatures, and during the Iran-Iraq war, they experienced an average of three small oil spills per week, which jeopardized their survival. Oil causes suffocation and respiratory failure of the dugong, and when detergents are used to dissipate the oil, the situation grows even worse because tiny oil globules enter deep into the respiratory tract and clog the tiny air passages. Oil also contaminates and damages the sea grass. As luck would have it, evidence in June 1991 suggested that no dolphins or dugongs died from the spill. These creatures were fortunate, but the whole sorry episode should serve as a dire warning for those who wish to wage war in the future.[49]

According to the World Wide Fund for Nature, the Persian Gulf is a major migratory flyway for millions of water birds, many of which are now being threatened. Information available four months after the war suggests that 30,000 to 60,000 birds have died. The spill also threatened some of the Gulf's major prawn and pearl fisheries.[50] Thousands of green turtles faced a catastrophic breeding failure as their nesting sites were cut off by the slick.[51] Although there were hardly any attempts to protect the environment from the oil spills, the International Maritime Organization spent $4 million of its own money to clear the prime turtle nesting areas, so very few turtles died.[52]

Two other environmental hazards were little discussed during and after the war. The Americans "took out" the small Iraqi nuclear reactors.[53] This action could well have caused the fissioning uranium fuel to melt down and release into the environment quantities of deadly radioactive poisons like two very small Chernobyls. What, if any, were the medical consequences?

The Americans also "took out" Iraqi's chemical weapons factories, which means that the chemicals in the weapons were almost certainly released into the air, because the chemical poisons do not disappear when the container of the bomb is damaged.[54] If a quantity of nerve gas less than the size of a dime contacts a person's skin, the victim dies within several minutes after experiencing seizures and respiratory paralysis. Mustard gas inflicts shocking burns on the skin and respiratory tract, causing severe breathing difficulties. How many people—innocent civilians—were exposed to these chemicals and have died as a result of the American bombing? There has been little if any reporting about these two issues.

The May 1991 issue of *Scientific American* reported that the Department of Energy released a memorandum to its numerous facilities saying,

> DOE Headquarters Public Affairs has requested that all DOE facilities and contractors immediately discontinue any further discussion of war-related research and issues with the media until further notice. The extent of what we are authorized to say about environmental impacts of fires/oil spills in the Middle East

follows: "Most independent studies and experts suggest that the catastrophic predictions in some recent news reports are exaggerated. We are currently reviewing the matter, but these predictions remain speculative and do not warrant any further comment at this time."[55]

This directive, it seems, still stood some months after the end of the war, when blazing oil wells continued to spew over 50,000 tons of sulfur dioxide (an acid rain producer) and 100,000 tons of soot a day.[56] The world must know the environmental consequences of the Persian Gulf war; the Department of Energy has no right to censor this devastating information for its own protection. At this point it is important to note that George Bush founded Zapata Oil; that Secretary of State James Baker is linked with Exxon, Mobil, Standard Oil, and Kerr-McGee; that Dan Quayle is backed by Standard Oil; and that the Department of Defense is the world's largest consumer of oil.[57]

Our addiction to oil is once again threatening the very foundation of the food chain and life-support system. It is so great that we must fight for oil, spill it, and, in the process, destroy even more of the planet. It took one million years to produce the quantity of fossil fuel that the global economy consumes each year.[58]

Nuclear meltdowns, poisoned gas, over one hundred thousand human deaths, dead birds, greenhouse gases, acid rain, toxic by-products from oil fires and new oil spills—all for oil. What will we do next? I am reminded of an alcoholic fighting for his next drink, smashing family treasures, hurting his children, killing his wife, and burning down his house.

SEWAGE

In my medical school days, we had to inspect the Adelaide sewage "farm" as part of our public health course. I remember feeling somewhat nauseated as I watched a man shoveling feces from a trench into a holding vat. When this material matured, it was used to fertilize fields and to grow food, after the proper elimination of pathogenic bacteria and parasites. China has for centuries been using human excreta as fertilizer. Sewage was a prized asset, and people who controlled the town sewage system were among society's elite. When I visited China in 1988, I was amazed by the extraordinary abundance of the crops.

Human waste is actually one of the most effective forms of fertilizer, and I have always wondered why we throw it away instead of recycling it. It comes from our food; it should go back to the food chain.

But today's sewage is polluted with toxic household chemicals and industrial waste. Apart from clean human waste and soap, nasties such as cleaners, bleaches, paints, heavy metals, pesticides, arsenates, cyanide, and radioactive isotopes join the waste stream. These chemicals obviously make sewage extremely dangerous.

If sewage is disgorged into the sea, many of these chemicals concentrate in fish. Sydney, with a population of three million people, has no sewage treatment facilities, and raw sewage contaminated with bacteria and chemicals is discharged into the ocean quite close to some of the most beautiful fishing and surfing beaches in the world. Sydney is a wonderful city—more beautiful, some say, than San Francisco—yet the state government is so short-sighted that it has blighted tourism by contaminating these beaches with raw sewage. Syringes, condoms, tampon applicators, plastic bags, and feces wash up on the golden sands.

The Mediterranean has been similarly contaminated. In 1990, several hundred dead dolphins washed up on the beaches of Majorca, the Costa Brava, and other Spanish resorts. Veterinarians concluded that toxic chemicals in the water had weakened their immune system. The azure blue waters of the Mediterranean have been terribly polluted over the last few decades by the enormous industrialization of southern Europe and by fifty million tourists on the beaches. The Rhone River alone, which passes some of France's most potent petrochemical and nuclear power plants, daily deposits into the sea an enormous amount of nitrates, sulfates, iron, sewage, mercury, and pesticides.[59]

People in New York and New Jersey were dismayed several years ago to find their beaches covered with garbage similar to that on Sydney beaches. Pollution has become universal.

Sewage contains large concentrations of phosphorus, an element used in detergents to soften water. Phosphates are potent plant fertilizers. When phosphate-rich sewage enters water systems, algal growth is promoted. Many waterways are clogged with algal blooms and overgrowth. The algae may totally cover the surface and, as they grow, absorb the dissolved oxygen in the water. The subsequent oxygen depletion causes fish kills. This process is called eutrophication.

Oceans are also contaminated with radioactive waste. Sellafield, a large nuclear reactor complex in Britain, has expelled huge quantities of radioactive elements into the Irish Sea, making it the most radioactive sea in the world. The elements include cesium 137, strontium 90, and plutonium 239. When released into the environment, these and many other radioisotopes concentrate thousands of times at each step of the food chain—from algae to crustaceans, to small fish, to big fish. It is almost certainly dangerous to eat fish caught in the Irish Sea. Not all fish will be radioactive, but fish swim thousands of miles, and it is impossible to detect which fish carry radioactivity in their flesh. When humans eat radioactive food, the elements are absorbed through the wall of the gut into the bloodstream, to be deposited in specific organs. Strontium 90, which is poisonous for six centuries, deposits in bone, cesium 137 (equally long-lived) settles in muscle, and plutonium 239 concentrates in bone, liver, spleen, and testicles and crosses the placenta into the developing fetus. Although these three elements are all potent carcinogens, plutonium is the most dangerous. One pound of plutonium could induce lung cancer in every human lung on earth if adequately distributed, and it remains radioactive and poisonous for half a million years. For example, if I die of a plutonium-induced cancer and am cremated, my smoke and the plutonium leave the crematorium chimney, to be inhaled by another human being; thus the cancer cycle continues ad infinitum. The latent period of carcinogenesis is five to fifty years—the period of time before cancer develops after exposure to radiation.

By now, hundreds of pounds of plutonium have entered the Irish Sea. Wherever in the world nuclear power plants are located, radioactive waste is discharged into seas, rivers, or lakes. All reactors need thousands of gallons a day for cooling, and this water is routinely flushed back into the water system, inevitably polluted with radioactive elements.

Another source of radioactive pollution of oceans is sunken nuclear submarines. Many of these ships carried not only nuclear reactors but also nuclear weapons, each containing ten pounds of plutonium. As the reactor and bomb containers rust, radioactive elements escape into the sea. Since 1945, twenty-seven submarines have sunk, including five Soviet, four U.S., three British, and three French.[60] There are at least 50 nuclear bombs on the ocean floor and eleven nuclear reactors that powered submarines, and there remain 15,000 to 16,000 nuclear weapons in the world's navies.[61]

Recently, the United States and Japan reached an agreement that allows Japan to reprocess its spent or used civilian nuclear fuel in France. Four hundred pounds of plutonium will then be shipped by air and sea back to Japan over a period of thirty years.[62] It should go without saying that the chances that a serious accident—one in which the highly lethal plutonium could be released—will occur during such a transport are high.

Now we move back to land to examine radioactive pollution created by the nuclear weapons industry over the last forty-five years. The Department of Energy (DOE) supervised the construction of 35,000 atomic and hydrogen bombs, contracting the work out to private corporations, including Rockwell International, Dow Chemical, General Electric, Westinghouse, Kerr McGee, and Du Pont. The building of seventeen huge nuclear facilities around the country proceeded in total secrecy in order to protect "national security." Hence neither the DOE nor the private contractors were held accountable for nuclear accidents, radioactive spills, nuclear meltdowns, leakages, and deliberate releases into air, water, and soil.[63]

Not only were tens of thousands of workers contaminated

over time, often without their knowledge, but hundreds of thousands of American citizens were irradiated. Some populations were actually used as guinea pigs and exposed to radiation as part of an experiment. Many people are now developing cancer, and their babies are being born deformed. Animals living near the nuclear weapons plants were also exposed to clouds of radioactive isotopes, and there have been clearly defined epidemics of deformities directly related in time to the release of radioactive materials. Radiation, in levels exceeding those at Chernobyl, was systematically released into the air and water at the Hanford bomb facility, in Washington State, more than forty years ago.[64]

A series of articles in the *New York Times* exposed these acts of negligence when the information first became available for public scrutiny, in 1988 and 1989, and I will quote from several of these stories. Remember that carcinogenic chemicals are synergistic biologically with radioactive isotopes in human and animal bodies.

In Hanford, Washington, a complex of nuclear reactors and 177 underground tanks containing high-level radioactive waste is located adjacent to the Columbia River. The tank complex is euphemistically called a "tank farm." The waste material contains over a hundred long-lived isotopes, including plutonium, cesium, and strontium, dissolved in concentrated nitric acid. These *must* be isolated from the environment for thousands of years.[65]

Not only is nitric acid extremely corrosive, but the radioactivity causes the liquid to boil at high temperatures. Amazingly, the DOE kept no record of the precise content and concentration of isotopes in the tanks. In the mid-1950s, potassium ferrocyanide was added to the tanks in order to concentrate the radioactive waste.[66] More recently, experts have recognized that cyanide combines with nitric acid to form an explosive combustible mixture and that it is therefore possible that the tanks will explode, scattering the deadly contents much as the Chernobyl explosion did. But many of the tanks are single-vessel steel, and between 1958 and 1975 twenty developed cracks and 430,000 gallons escaped. This material is migrating toward the Columbia River.[67] In all, 700,000 to 900,000 gallons have leaked.[68] Twenty-seven streams around Hanford are contaminated with radiation, and rabbits and coyotes using the radioactive waste as salt licks have scattered radioactive dung over an area of two thousand acres.[69]

Until 1971, the nuclear reactors were cooled with water from the Columbia River, and the radioactive coolant was discharged back into the river on a daily basis. From 1945 to 1971, tens of millions of curies (a curie is a measurement of radiation) were dumped into the river. In the early 1960s, a Hanford worker ate oysters caught hundreds of miles downstream at the mouth of the Columbia River. When he went to work the next day, he set off the radiation alarm at the Hanford plant.[70] The Columbia River is a rich spawning area for salmon, and I have seen fishermen in Astoria at the mouth of the river catching salmon. Why are they not warned about the dangers, and how many thousands of other people now radioactive are not checked by radiation monitors?

In September 1991, the General Accounting Office announced an investigation into the mysterious disappearance in 1989 of key documents that calculated leaks from the radioactive waste tanks. Apparently, 750,000 gallons of cooling water were pumped into a leaking tank, and 750,000 gallons of radioactive liquid escaped. The GAO also reported that 444 billion gallons of liquid emanating from reactors, processing plants, and tanks were discharged into the environment at Hanford between 1945 and 1991.[71]

Not only has the Hanford plant been discharging and leaking radiation into the river for forty-five years, but serious accidents have occurred at the reactors. One could perhaps excuse an accidental release of radiation, but on several occasions huge clouds of isotopes were created knowingly and willfully. In December 1942, about 7,800 curies of radioactive iodine 131 were deliberately released in an experiment designed to detect military reactors in the Soviet Union (only 15 to 24 curies of iodine 131 escaped at Three Mile Island in 1979). Then, between 1944 and 1955, over half a million curies of iodine 131 were released. This element concentrates in the food chain, particularly in cow's milk and human milk. Babies and young children are ten to twenty times more sensitive than adults, because duplicating genes in actively dividing growing cells are particularly vulnerable to damage from ionizing radiation; and iodine specifically concentrates in the thyroid gland. An abnormally high incidence of thyroid tumors and cancers has been observed in populations living downwind from Hanford. Strontium 90, cesium 137, and plutonium 239 have also been released in large quantities, as was, between 1952 to 1967, ruthenium 106. People in adjacent neighborhoods were kept uninformed about these releases—before, during, and after—and none of them were warned that they were at risk for subsequent development of cancer.[72] (Some experts have estimated that downwind farms and families received radiation doses ten times higher than those that reached the Soviet people living near Chernobyl.)[73]

A thirty-seven-year-old uranium milling plant at Fernald, Ohio, near Cincinnati, has since 1951 released 298,000 pounds of powdered uranium into the air and 167,000 pounds into the Great Miami River; another 12.7 million pounds of it discarded into earthen pits.[74] Uranium, the fuel for nuclear reactors, is also used as casings for hydrogen bombs. When fissioned in a nuclear reactor, it is converted into more than two hundred isotopes, all biologically more dangerous than uranium itself. It is nevertheless a potent carcinogen in its own right, one that, if swallowed, migrates from gut to bone. Several years ago, a young boy living near the plant developed osteogenic sarcoma (bone cancer). After his leg was amputated, the bone was analyzed and found to be contaminated with high concentrations of uranium. Incidentally, the uranium milling plant was for many years disguised from the local population, for it was called the "feed materials production center," masquerading as a pet food factory.

Other radioactive weapons plants are located at Oak Ridge, Tennessee, at Rocky Flats, Colorado, on the Savannah River, South Carolina, at West Valley, New York, and at Idaho Falls. The DOE recently for the first time admitted that hundreds of thousands of people living around these diabolical facilities have been exposed to accidental "releases" since 1945. Similar stories of tragedy and grief are common among all these populations.[75]

Because of an unusual cluster of primary brain cancers in Los Alamos, the New Mexico Department of Health and the U.S. Department of Energy have initiated a study to determine how much radiation the Los Alamos Nuclear Weapons Laboratory has released into the environment since 1943 and to assess whether the incidence of brain cancer in the total community of eighteen thousand people is excessive. Over the years, the laboratory released millions of gallons of radioactive and toxic wastes, which were dumped secretly into the nearby canyons and ravines. But the DOE has no official documentation regarding the quantities dumped or even the location of the dump sites.[76]

In the 1960s, targeters at the Pentagon told the DOE to cease manufacture of bombs, because they had all relevant targets in the Soviet Union covered, but the bombs continued to be produced, as if by some sort of gigantic, autonomous, uncontrollable production line monster. The cold war is over, the Berlin wall is down, but the DOE was until recently still turning out up to five new hydrogen bombs every day.

In 1989, Congress appropriated $240 billion to "clean up" fifteen of these weapons-producing sites. The problem is the same as at Love Canal: how does one retrieve radioactive elements long since scattered to the four winds, hiding in dead and living bodies (animal and human), migrating from earthen pits, leaching into underground aquifers and rivers, and concentrating forevermore in radioactive food chains?[77] *C'est impossible!* Apart from the radioactive waste at these weapons sites, there are nearly four thousand poisonous-solid-waste sites—no one knows exactly how many, because record keeping by the DOE and its contractors has over the years been bad.[78] Although it has a thirty-year plan to "clean up" these radioactive sites, the DOE admits that it is a next to impossible task—when you dig the waste up, where do you put it?[79]

Catastrophes at nuclear weapons plants reflect similar ones at civilian nuclear reactors, some leaking radiation and all creating massive quantities of radioactive waste on a daily basis. There are 111 active plants in the United States and 430 reactors worldwide. The Nuclear Regulatory Commission has estimated that there is a 50 percent chance that a meltdown larger than Chernobyl's will occur in the United States by the year 2000.[80] Because radiation knows no boundaries, the consequences of a meltdown affect many countries. Fallout from Chernobyl spread not just to the Soviet Union but also large areas of Europe extending through Poland, Austria, Germany, northern Italy, France, the Scandinavian countries, and parts of Britain.

An article in June 1990 predicted that 800,000 children in the Soviet Union could be at risk for developing leukemia in the wake of the Chernobyl accident. It also reported that many babies have been born without arms or legs or with other gross deformities. Doctors estimate that about 160,000 children below the age of seven years living in the Ukraine in radioactively contaminated areas are at risk for developing thyroid cancer (from radioactive iodine 131). Another 12,000 are also at risk because they drank contaminated milk and inhaled radioactive iodine before they were evacuated from the thirty-kilometer zone around the reactor. In Belorussia alone, 2,697 villages, with a combined population of two million people, have been seriously contaminated. People in the Soviet Union are still eating radioactive food (food grown in contaminated areas will be radioactive for thousands of years). Some Soviet doctors and environmentalists have said that a total of 3.5 million people are at risk for cancer or leukemia.[81] Soviet scientists call the radiation sickness, and associated cancer and leukemia, Chernobyl AIDS, because the disease bears a close medical resemblance to AIDS.

On the fifth anniversary of the Chernobyl disaster, I was interviewed on the radio along with Robert Gale, the doctor who treated many of the acute radiation victims, and John Gofman, a world-famous expert in radiation. It was clear that it continues to be a tragedy, and doctors are still not sure of the true medical consequences. Leukemia takes five to ten years to become clinically apparent and solid cancers fifteen to sixty years. Some leukemia cases are beginning to be diagnosed.

Despite this uncertainty, new data at this time revealed that approximately 650,000 people participated in the cleanup operation after the accident and that these people, called liquidators, were exposed to very high levels of radiation. According to various estimates, between 5,000 and 10,000 of them have already died from radiation-induced illnesses. Five million people (one-quarter of them children) still live in contaminated areas five years after the accident and 300,000 in highly contaminated places.[82] The Soviet government does not have the facilities or the resources to relocate these people or adequate medical equipment to monitor the internal radiation levels, let alone to provide bone marrow transplants and medical treatment for the many expected leukemia victims of the future.

Although there are no definitive scientific answers to the most serious problem of radioactive waste, certain issues which must be addressed immediately are:

- All nuclear reactors, both military and civilian, must be closed down at once so that no more waste is produced, no more bombs are made, and no more accidents occur.
- Two-thirds of the scientists in the United States work for the military-industrial complex. As the cold war is over and war is clearly obsolete, these brains should be transferred from weapons manufacture to the urgent task of finding safe radioactive-waste storage. We did not inherit the earth from our ancestors; we borrowed it from our descendants. If we fail to solve this dreadful problem, radioactive food and water will be our legacy to future generations, with increased incidences of genetic disease, deformed babies, and epidemics of children dying of cancer and leukemia.
- The global stockpile of 60,000 nuclear weapons should be immediately dismantled and the plutonium stored along with the nuclear waste. Each weapon contains 10 pounds of plutonium; in total, 250 tons of this metal are contained in the nuclear stockpile and 250 tons in America's radioactive waste. Plutonium must never be allowed to enter the ecosphere, although large amounts have already been released at many places, including Rocky Flats, Colorado,

contaminating Denver and surrounding reservoirs and suburbs.

- A concerted effort must be made to retrieve the spilled and leaked radioactive waste, and a suitable, safe disposal must be made available. But it is quite unlikely that radioactive waste poisonous for hundreds of thousands of years can be isolated from the environment for a hundred years, let alone for lengths of time spanning earthquakes, glaciers, and other geological disturbances.

To summarize this chapter on chemical and radioactive poisons: we must without delay educate ourselves about the state of the planet, we must alter our life-styles so that we do no more harm, we must not allow enormous quantities of chemicals to be manufactured, we must close down polluting industries, we must fine and jail polluters who persist, and we must clean up our planet.

5

Species Extinction

> In my book, a "pioneer" is a man who turned all the grass upside down, strung barbed wire over the dust that was left, poisoned the water and cut down the trees, killed the Indian who owned the land and called it progress. If I had my way, the land would be like God made it, and none of you sons of bitches would be here at all.
>
> —Charlie Russell, cowboy artist, 1923

We have taken over the planet as if we owned it, and we call it progress because we think we are making it better, but in fact we are regressing. Species are dying in the wake of this "progress," and we seem not to realize that our life depends upon theirs. Peter Raven, director of the Missouri Botanical Gardens, in St. Louis, says that the destruction of species is more critical for the world than the greenhouse effect and ozone depletion, because it is moving faster and is inevitable. He predicts that over the next thirty years human beings will cause the extinction of a hundred species per day.[1] For fifteen years, I traveled the world warning people about the medical and ecological consequences of nuclear war, not aware that life was already dying quietly and unobtrusively from man's ongoing activities. Now I see that the threat of species extinction is as serious as the threat of nuclear war.

Life began on the planet 4 billion years ago, and over time an astounding array of diverse forms have gradually emerged. But the advance of evolution has not always been smooth. There was a rapid increase in the numbers of species up to 600 million years ago, with a subsequent decrease in diversification for the next 200 million years. In the last 400 million years, the numbers slowly increased, with the interruption of five significant extinction phases. The largest of these was the Permian era, 240 million years ago, when 77 to 96 percent of all marine animal species became extinct. It took another 5 million years for species diversity to recover. About 65 million years ago, another significant era of species extinction began when the dinosaurs, which had previously ruled the earth, disappeared and mammals gained global hegemony. Thus the evolutionary stage was set for the appearance of *Homo sapiens*. Since that time, the numbers of species have continued to increase to the present, all-time high.[2]

Human beings first appeared in a primitive form some three million years ago. That species lived in relative harmony with other life forms until only ten thousand years ago, when it began to have a devastating effect upon the diversity of other species. Humans hunted animals and birds and chopped down or burned forests and plants. To give two examples, in Polynesia, in an isolated island environment, one-half of the bird species are extinct because of human hunting and forest destruction, and more recently, during the 1800s, almost all the unique shrubs and trees were destroyed on the small island of St. Helena, in the North Atlantic. Of the 30 million species estimated to be extant today, we may now be losing 17,500 each year.[3]

In this time frame, the span of *Homo sapiens*'s existence seems trivial. Yet we now threaten to exterminate most of the world's species, which have taken four billion years to evolve. Although the dinosaurs disappeared, they did not, as we may well be doing, bring about their own extinction.

Unlike the dinosaurs, we clearly have almost total dominion and control over the planet. The development of the opposing thumb gave us the ability to make and hold instruments, weapons, and tools of mass destruction. Our abnormally large neocortex—the thinking part of the brain, which developed in a very short evolutionary time frame—has enabled us to communicate thoughts, by speech and writing, and to destroy, dominate, and subdue all other species, for our own benefit, as we perceive it. We are shortsighted and egocentric, little realizing that our survival is intimately related to and depends upon the survival of 30 million other species. This behavior may, in the long run, be a kind of suicide. The behavior of most other species is conditioned for their long-term survival. For example, humans stand at the apex of the pyramid of the food chain. Bacteria in the soil break down the fallen leaves to produce humus and compost to feed the plants to feed us. If we kill the bacteria with chemical toxins and change of climate, we will indirectly kill ourselves. We have a similar relationship with many other species.

We really "came into our own" with the dawn of the industrial age, early in the 1800s. As we harnessed nature and worked with the natural laws of science, we learned to destroy forests and pollute the air, water, and soil very efficiently, and this efficiency has, over the last two centuries, increased exponentially. In the grand scheme of evolution, our obsession with interpersonal, national, and religious conflicts and our wars are meaningless.

Although 30 million species of plants and animals have not been systematically documented, scientists arrive at this number by extrapolation, having analyzed the number of new species in a small area of rain forest. Within each of these 30 million species, there is a huge degree of genetic diversity that is terribly important. Let me use the species of *Homo sapiens* as an example. Each person carries a unique set of genes that was derived from a particular sperm that fertilized a particular egg. Each sperm and egg was genetically different from any other sperm or egg, and this unique gene structure (genotype) determines all the facial and bodily characteristics, personality, and mental abilities of an individual. The process of evolution depends upon genetic diversity, because as the environment changes, only those organisms with specific characteristics that allow them to survive the change will reproduce. This is what Charles Darwin called survival of the fittest. Without genetic diversity, evolution could not have happened. To give a simple example, some breeds of maize are able to survive better in climatically unfavorable conditions than are others. Scientists are finding that by mating one variety of wild maize with a domestic type, a better cropping plant can be produced. So instead of concentrating on saving a few individuals within a particular species, we must save all the variants of life forms within each species.

There are thousands or even millions of varieties of plants that we have not yet even identified, but as we destroy the environment, we will be needing special strains of wild maize, wheat, rice, and other species that will grow in difficult terrains and climates. Over thousands of years, the human race has utilized about 7,000 different plant species for food, but the present generation tends to rely upon only about 20 species to provide 80 percent of the world's food. These 20 include rice, wheat, millet, and maize.[4]

We consume less than 0.1 percent of naturally occurring species. But we do know that more than 75,000 plant species are edible and that some are far more appropriate than those we now use. Edward O. Wilson has described a plant called the winged bean, or *Psophocarpus tetragonolobus,* from New Guinea, which is at present ignored by the world's food manufacturers and farmers. The whole plant is edible—roots, seeds, flowers, stems, and leaves—and a coffeelike drink can be made from the juice. It grows rapidly, to a height of fifteen feet in several weeks, and exhibits a nutritional value equivalent to that of the soy bean.[5]

At a time when the human population is growing explosively and needs an enormous amount of food, it seems imperative that we start preserving and cultivating different plant and food varieties that will provide efficient sources of nutrition.

Ironically, as other species become extinct, we are proliferating. In 1800, we numbered 1 billion; in 1990, 5.2 billion. Over the next ten years, we will add another 1 billion. By the end of the next century, if present trends continue, we may reach 14 billion, or three times our present number.[6] Clearly, the ecosphere cannot sustain 5.2 billion, let alone 14 billion. (Overpopulation is discussed in chapter 6.)

Let me now introduce the varieties of ways in which species are being destroyed.

TROPICAL FORESTS As was noted in chapter 3, these special ecosystems probably contain 15 to 24 million of the 30 million planetary species. Within twenty-five to fifty years, the trees and those species may well be gone, destroyed because of Third World debt and First World greed.

WETLANDS These serve as crucial breeding grounds for fish, crustaceans, and other sea-dwelling creatures. Until very recently, though, mangrove swamps and reedy wetlands were not valued as habitats for species. Rather, they were seen as ugly, muddy, difficult areas that were best cleared for "canal developments" or filled in for real estate investment. Although some people now understand the ecological significance of wetlands, most "developers" are still uneducated, and indeed the majority of people in the Western world are deeply ignorant about the biological meaning of species extinction. Some 25 to 50 percent of the global wetlands have so far been destroyed. In the United States, this number is 54 percent.[7] As the human population increases, it decreases the number of fish by eliminating their breeding nurseries. Yet fish are a key source of first-class protein for millions of people.

CORAL REEFS These are a rich source of species and are very complex ecosystems in their own right. An area of 400,000 square kilometers of coral reef is estimated to contain 500,000 species. Many of these animals and fish engage in a kind of biological warfare with one another, and they have therefore evolved a large number of specific toxins that can be harnessed for medical treatment and possibly other advantageous uses. But coral reefs are mysteriously dying all around the world. It seems that coral polyps (the living organisms that provide the vivid colors and that create the solid coral structure as their protective habitat) are very sensitive to temperature change, so their death could signify early global warming from the greenhouse effect. Worldwide temperatures in the 1980s were 4°F above those of any previous decade on record. Toxic chemicals derived from agriculture, industrial waste discharged into rivers, and urban runoff from houses (pesticides, cleaners, fertilizers, and sewage) could also be killing the coral. These magnificent reefs, true wonders of nature, must be preserved at all costs, and no change in climate or human "development" should be allowed to threaten their existence.[8]

ARID ZONES OR DESERTS These are habitats to many fragile species of plants and animals. Australian deserts were once filled with tiny exotic marsupials that looked like delicate, forlorn, scaled-down kangaroos, as well as other unique and precious life forms.[9] Of the seventy-two different mammalian species that once lived in the desert, eleven are now extinct, five are found only on isolated "safe" islands, and fifteen are threatened. These indigenous animals were decimated by the introduction of foreign species of what are called feral animals. Foxes prey on small animals, as do wild cats and dogs, while grass-eating rabbits, pigs, goats, camels, buffalo, rodents, wild horses (called brumbies), and cattle have destroyed thousands of square miles of fragile plant ecosystems.[10]

INTRODUCED SPECIES I have just described the devastation wrought by introduced species that have no natural predators. But sometimes foreign animals are brought into a country specifically to prey upon a natural pest that needs to be eradicated. For instance, the South American cane toad was brought to Australia in order to eradicate a beetle that was destroying sugarcane crops in Queensland. Not only did it not kill the beetle, which lived out of reach on tall sugarcanes, but, having no natural predators, it has multiplied out of control and spread like a creeping plague over much of the east coast of Australia. I live on the east coast hundreds of miles south of the point of introduction, and at night my paths and gardens are alive with silent, hopping, slimy cane toads, which are brown, warty, and ugly. They not only look repulsive but have poisonous glands located in the area of their head. If they are eaten by animals such as snakes, goannas, lizards, cats, or dogs, the animal is poisoned and dies. The cane toad plague is now so serious that many of our indigenous reptiles and other creatures are threatened with extinction. (I may sound somewhat judgmental in regard to the cane toad, but I would have none of these feelings if it were located in its original environment.)

SAVANNAS AND GRASSLANDS The wildlife of these regions is also under threat, particularly in Africa, because human beings enjoy displaying exotic animal skins on their floors, hanging animal heads on their walls, grinding rhinoceros horns for aphrodisiacs, wearing shoes and belts made from crocodiles and snakes, and making jewelry and piano keys from elephant tusks.

In Africa in the 1930s, there were ten million elephants; now there are fewer than three-quarters of a million. Eighty thousand elephants are killed each year by poachers who hack off their tusks with chain saws and leave the huge carcass to rot in the midday sun, while they make a living from their illegal bounty.[11] These poachers are often poor, and because the world supports an unequal distribution of wealth, this is their one mode of survival, the destruction of nature. Lions, tigers, leopards, zebras, giraffes, and other wonderful exotic animals are all endangered. Soon, these animals will exist only in zoos, and eventually these few remaining specimens will become extinct.

The total population of other creatures is also diminishing. For instance, blue whales have decreased in number from 200,000 to 15,000, and humpback whales from 50,000 to 8,000.[12] The number of southern white rhinoceroses in Zaire dwindled from 400 in the 1970s to 15 in 1984; the population is now up to 28.[13]

WILDLIFE SMUGGLERS There exists a very lucrative international trade in wild animals and their component parts. As a member of the South Australian National Parks and Wildlife Council in the 1970s, I learned how our beautiful indigenous cockatoos were being drugged and smuggled in socks packed in suitcases aboard international flights. Australian snakes, birds, and marsupials are also part of this international commerce.

On a trip to Crete in 1987, I was walking along a side street when I heard a familiar scream—it was a sulfur crested cockatoo looking very bedraggled and dirty in a cage outside someone's shop. Annoyed to see this smuggled bird, I grew even more incensed when I realized that it should by rights be flying free with a flock of hundreds of others in the Australian bush. How much did its captors pay for it? In all likelihood, $2,500 on the black market.

The international trade in wildlife and its products is now worth $4 billion to $5 billion a year (excluding fish and timber). Bangkok is used as an international transit station by dealers who "launder" illegally obtained wildlife, which is poached in Indonesia, Laos, Vietnam, and Cambodia. This international Mafia is threatening the extinction of the great panda, crocodiles, alligators, snakes, cacti, and orchids. Recently, a traveler from Mali was intercepted in Paris, and his luggage was found to contain fifty pythons, twenty tortoises, twenty lizards, and several vipers. A Japanese tourist arriving in Bangkok had eleven rare monkeys jammed into a carry-on bag, five of them had suffocated. Monkeys' teeth are often extracted with pliers and cut with clippers to make their bites harmless, and leopards' fur is dyed black to make it look like a house cat's. Americans have created a huge demand in rare parrots and dangerous snakes, while the Japanese like monkeys.[14]

CHEMICAL DESTRUCTION OF WILDLIFE Some of the most important, and yet seemingly insignificant, species in the world are threatened by toxic chemical sprays used on crops. These creatures are the worms, fungi, insects, and bacteria that maintain a healthy soil base and root system for plants. They form the base of the pyramid of the food chain. Bees and other insects that pollinate crops and disperse seeds are also undervalued but are vital to our survival. Bats, which represent one-quarter of all mammalian species, also scatter seeds. We probably do not realize that 90 percent of the most valuable U.S. crops, worth a total of $4 billion, are fertilized by insects, and the catch-22 is obvious. Pesticides and weedicides used to protect the crops from predatory insects and weeds kill the very organisms upon which the crops depend.[15]

Wild birds, bats, and parasitic insects have another function. They eat insect pests and therefore act as natural insecticides. So nature is clearly best left to itself. It has all the inbuilt mechanisms and feedback loops to ensure its ongoing health and survival. To put it crudely, we humans just screw things up. Organic natural farming is the only sensible alternative to chemically destructive farming.

THE DOMINO EFFECT If we upset the balance of nature by eliminating large predatory animals, we produce a reactionary overpopulation of smaller predators that are normally kept in check. These in turn then eat and render extinct the lower-order animals upon which they feed. We must not disturb the hierarchical balance of nature and the food chain.

FROGS—THE NEW CANARIES Years ago, canaries were used as sensitive indicators in coal mines to determine whether the air was safe for the miners to breathe. When the canary died from toxic fumes, it was time for the miners to leave. For the last decade, biologists have noticed an alarming decline in the numbers and species of amphibians—frogs, toads, and salamanders. These creatures were the first vertebrates ever to inhabit the land. They appeared 400 million years ago and evolved into present-day species some 200 million years ago. Because their bodies are acutely sensitive to the environment, they are our canaries, our barometer of global environmental poisoning.

Frog species are disappearing from Australia, the United States, Japan, Canada, Puerto Rico, and Costa Rica. This phenomenon is particularly worrisome because they are disappearing from national parks and from some of the best-protected areas of the world, where they should be safe.

What is threatening them? Many theories have been proposed: *(a)* frogs' eggs are very sensitive to UV light; *(b)* foreign predator fish introduced into frog habitats, such as ponds and lakes, eat the tadpoles and baby frogs; *(c)* toxic chemicals and heavy metals are poisoning the frogs because frogs naturally absorb large amounts of water, some of which is now polluted, through their skin; *(d)* habitat damage caused by logging, pesticide pollution, and dam construction threatens frogs' survival; *(e)* frogs are so sensitive to variations in temperature and moisture that if the rainfall and climate change, the frogs die—and these changes seem to be occurring as the greenhouse effect becomes manifest.

Frogs are an important link in the ecological chain. Tadpoles consume large quantities of algae, so streams are kept clean and flowing, and adult frogs eat enormous numbers of insects, including mosquitoes. The Everglades, in Florida, offer a good example of the ecological necessity of frogs. Scientists recently recorded a 90 percent reduction in the number of wading birds, possibly caused by the demise of pig frogs, which provided the birds' food.[16] If frogs are the new canaries, the situation is very serious and we had better act fast.

DESTRUCTION OF HABITATS BY TRAPPING AND MINING The Gouldian finch is an exquisitely colored bird once found throughout the grassy woodlands of tropical Australia. Fifty years ago, flocks of thousands used to swarm through the air, but the swarming has stopped. Now flocks of only fifty or fewer are seen, because the finch was extensively trapped for the birdcage industry and because its habitat was destroyed by the regular, intentional burning of the grasslands. The bird nests were vulnerable to destruction since they were located in termite and ant mounds close to the ground, or in the hollows of old gum trees. A recent survey found only two intact breeding colonies in the whole country, and one of these is at present threatened by a gold-mining company. The company plans to mine at the site of the breeding colony and to construct tailings dams close by. These dams will hold cyanide-contaminated water, so when the birds drink their daily fill of water, they will die. Is this the epitaph of the Gouldian finch?[17]

ANTARCTICA One of the last bastions of pristine wilderness is now under threat. The Antarctic is an ecosystem delicately balanced and teeming with life. No land plants grow on the icy waste, but the sea supports all the life forms. The base of the food chain consists of single-celled plants called phytoplankton, which trap energy from the sun. Billions of tiny crustaceans known as krill eat the plankton, which in turn forms the food base of whales, penguins, crabeater seals, and large seabirds such as the petrel. Small fish and squid also eat the krill and in turn feed emperor penguins, albatrosses, large fish, seals, and sperm whales.[18]

Although the climate of Antarctica is not conducive to human habitation, we have nevertheless devised many ways to intrude and damage this unique and fragile biosphere:

- Large-scale international commercial fishing is depleting the sea.
- In February 1989, an oil spill from an Argentinean tanker killed thousands of penguins, skuas, and their chicks.
- The tourist industry recently decided that the habitat of the South Pole is a nice place to take its customers. They leave a trail of desolation as tons of garbage are dumped from tourist ships into the sea. Incidentally, most ship captains in the world still believe that the oceans are a universal sewage disposal system and act accordingly.
- For fifty years, scientists established research stations in the Antarctic, but they did not build adequate sewage systems and left hundreds of cans of toxic waste and garbage when they departed. How could they research the delicate web of life and then so insensitively threaten it, or were they interested only in their research papers?
- Ozone destruction is an extraordinary threat. Remember that in the winter the ozone over the South Pole decreases to only 6 percent of its original 100 percent. Remember, too, that life cannot exist without the ozone layer. Plankton is extremely sensitive to UV light. As it dies, so will the rest of the life cycle.
- Mining corporations have been pressuring their governments for the right to dig up minerals in the Antarctic. That activity would lead to the ecological devastation of much of the area. Accordingly, an international treaty called the

Convention on the Regulation of Antarctic Marine Resource Activities has been drafted and was agreed upon by thirty-three nations in Wellington in June 1988. Fortunately, Australia, Britain, and, probably, France have decided not to sign it. The treaty will not be valid unless all nations agree to it. Since then, Australia has proposed that the world declare the Antarctic a park to be left in its original, pristine state. This may well be the favorable outcome.[19]

THE FISH AND THE SEA World market forces have helped destroy the ecology of the sea. New efficient techniques and excessive fishing have so depleted some fish species that they may never recover. Market forces are also responsible for pollution of the oceans by toxic sewage and poisons, plastic disposal, radioactive pollution, acid rain, and oil spills. Much of the oceanic life is still a mystery to us, because we cannot really explore the deepest layers of the oceans. Strange and wonderful life forms have, however, been dredged from great depths, but even these species are not immune to the danger of sunken nuclear submarines and land-based poisons that we tip into the sea.

Pollution of the sea by plastic kills large numbers of marine animals. The fish eat pieces of indigestible Styrofoam and plastic, which causes intestinal obstruction or blockage of the bowel, and they die a slow death from starvation. Seabirds also eat plastic, because it resembles fish, and they suffer similar deaths. Birds and fish often get caught up and trapped in the conjoined plastic rings that hold together a six-pack of beer or soda cans. These rings strangle birds and fish. I once found a dead albatross with one of these disposable obscenities wrapped around its neck, on the ninety-mile beach in South Australia. What a symbol of the industrial consumerist age! The National Academy of Sciences estimated that up to two million birds, ten thousand sea mammals, and countless fish die each year in American waters because of internal damage caused by plastics.[20]

Fishing ships have become fishing factories. Instead of catching tuna with rods, fishers now use huge seining nets to scoop up tuna and all other associated fish from large sea volumes. (Thousands of dolphins are also caught in these nets and die unnecessarily.) International quotas must be placed upon certain fish species in order to protect them. Tuna, for example, should be regarded as a luxury item and harvested accordingly, in relatively small numbers.

The Japanese and Taiwanese, whose diets consist almost solely of fish and rice, are profligate fishers. At present, they deploy drift nets made of finely meshed, colorless, invisible nylon sixty kilometers long that hang fifteen meters below the surface of the sea. The nets catch anything swimming in either direction along this barrier. They also trap seabirds that dive into the water to catch fish, and they ensnare whales, dolphins, sharks, turtles, and seals as well as the target fish, salmon and tuna. U.S. environmentalists estimate that more than three-quarters of a million seabirds are killed in drift nets each season, plus tens of thousands of dolphins.[21]

These nets in effect are strip-mining the seas. In 1989, they netted 20,000 to 40,000 tons of tuna in the South Pacific, instead of the usual 10,000 to 15,000 tons. As a result, albacore and bluefin tuna stocks of the South Pacific are threatened, and it is thought that many species will never recover.[22]

Many nations of the world have registered outrage at these fishing practices. As a result of international pressure, Japan will now stop drift netting a year earlier than expected, but Taiwan will continue to drift net despite the protests.[23] Japan, it must be said, hopes to resume drift netting in the near future, if it can negotiate certain points with South Pacific countries. So each victory becomes, in effect, only a holding action, until new efforts are made. Greenpeace has estimated that 6,400 dolphins and thousands of other fish of no commercial value were killed during a three-month drift net fishing season in the Tasman Sea.[24]

WHALES The Japanese are still whaling despite an international law that bans the killing of whales. This country continues to catch 1,800 whales per year for "scientific purposes." Their carcasses are cut up and sold for meat for sushi, sushimi, and other delicacies. However, whale meat in fact provides less than 1 percent of the protein in Japan. Since the international ban in 1985, the Japanese have killed 13,650 whales.[25] The blue, sperm, and right whales have been hunted to commercial extinction, and the fin and minke whales are threatened.

The Soviets stopped whaling in 1987, but they announced that they would take thirty fin whales and seventy minke whales a year in 1990 from the Okhotsk Sea—to broaden their "knowledge and understanding of the marine ecosystem." That means they will kill them to learn that they are becoming extinct.

DOLPHINS Dolphins are highly intelligent, having very large brains, and capable of communicating with human beings. They can count, repeat words in a primitive phonetic form, and communicate emotionally when they have befriended certain people. In addition to being slaughtered as a side effect of the large-scale fishing industry, they are dying spontaneously of mysterious causes, which may well be related to ocean pollution. For instance, scientists reported that an examination of 260 dead dolphins washed up on the beaches in Spain in August 1990 revealed that they were suffering from a viral infection similar to distemper in dogs. But they also had liver lesions caused by toxic substances in their blood that apparently entered their bodies from the contaminated water. Scientists believe that poisons in the sea are inhibiting their immune system, thus making them more susceptible to viral infections. This mechanism is very similar to the pathophysiology of AIDS, where the immune system is depleted by the AIDS virus and patients die of massive bacterial or viral infections and cancer.[26]

Tourists are encouraged to feed dolphins with fish on certain organized cruises in the States, but this practice has been discouraged by the Center for Marine Conservation, because the fish themselves may be infected.[27] Bottle-nosed dolphins and sea lions have been used for sinister purposes. The Reagan administration spent $30 million in a clandestine program to train these highly intelligent animals to guard the Trident submarine base at Bangor, Washington. While navy officials said that the dolphins and sea lions are somewhat unreliable during training and occasionally go "absent without leave or refuse to obey orders," they admitted that what the animals lack in discipline, they make up for in sonar and speed: "their sonar system is better than any radar and they can pick up objects with incredible accuracy." The Naval Ocean Systems Center, in San Diego, is a sort of "boot camp" for a hundred dolphins and twenty-five California sea lions.[28] Funding for this program has been cut, but the training program continues.[29]

SEA TURTLES These creatures are under threat in the South Pacific. Fiji exports turtle shells, and 2,025 kilos of shell are exported to Japan each year. This figure represents a total of two thousand hawksbill turtles. The trouble is that only the large, mature turtles are captured and killed, and since these mature creatures are the breeders, the future of the whole population is in jeopardy. Turtles live for more than a hundred years. Five of the six sea turtle species are in serious danger of becoming extinct, according to the World Wildlife Fund.[30]

Turtles are fascinating creatures. William, my son, and I stayed on Heron Island, on the Great Barrier Reef in Queensland, several years ago. As we lay luxuriating on the sand, huge female turtles clambered up the beach from the sea, laboriously dug a hole in the warm sand with their flippers, and proceeded to lay several hundred eggs. When the eggs hatched at night, the tiny baby turtles dug their way through the sand and headed toward the nearest light, mistaking it for the moon reflected in the sea. In this case, it was toward street lights, and the baby turtles became lost. In the morning, the beach was still alive with turtles, but the sea gulls snapped them up. Some eventually got to the water, where big fish waited to devour them. Thus, out of thousands of hatchlings, few survive. When they disappear into the ocean, no one knows where they go. They reappear years later as giant adults.

Before my trip to the Amazon in 1989, I visited the Galápagos Islands, which are cared for and protected by the Ecuadoran government. This group of islands is situated on the equator, several hundred miles west of Ecuador. The islands intersect about five ocean currents, which not only influence the climate but also have played a crucial role in the evolution of life forms on the islands. Although I expected the weather to be very tropical and hot, it was in fact tempered by the cool ocean current heading north along the coast of South America and the current heading south from the west coast of Canada and the United States. These currents had, over time, carried an interesting variety of wildlife from various parts of the world to the Galápagos.

On one of his voyages, Charles Darwin visited the islands, where he developed his ideas on evolution by noticing that a particular finch on one island had features subtly different from those of the same species on another island. These landmasses are just far enough apart that the birds could not fly from one to the other; hence they evolved different characteristics in isolation from one another.

We traveled nocturnally by boat from island to island, and each morning we awoke to a different exotic landscape. The islands are volcanic in origin, and one of them is indeed still an active volcano. Some consist of stark, barren basalt, like the surface of the moon, and most have no water supply. Plants are primitive and sparse, and tall cacti have evolved to escape the predatory giant land turtles. Because of the inhospitable landscape, humans have not inhabited many of the islands and hence have made little impact upon the wildlife. Giant land turtles are unique to the place, having migrated over eons from the sea to the higher ground and learned to graze upon grass. During the last century, pirates used to fill their holds with live turtles, which could survive for a year without food, and they would then be assured of a fresh, wholesome supply of meat. Now all the wildlife of the islands is protected, and I have never seen a region so teeming with life. Literally millions of birds, nests, babies, and eggs covered some islands. We walked along narrow paths snaking between nests built sometimes right on the path, and the birds just stood and watched as we walked. Red-footed boobies mated while blue-footed boobies stood in their nests looking as if they had blue rubber feet. I walked up the side of a cliff and came face to face with an owl that stared at me without expression. We came across a pair of albatrosses doing a ritualistic mating dance. These birds are monogamous, mating with one partner for life, but they court for about a year. The dance was beautiful to behold. They clacked their long beaks together and then waved their heads around in a lovely, sinuous movement. The event continued for hours. The birds are so big that they cannot fly without taking off down a self-made runway and launching themselves off the edge of a cliff.

Thousands of sea lions lazed like slugs over the sands. Big males guarded harems of one hundred females, while live babies, dead babies, and placentas were strewn across the sand. William lay on the beach to take a nap and awoke to look up into the faces of four sea lions that were scrutinizing him. Herds of smelly sea iguanas sunned themselves on black basalt rocks, and huge land turtles trundled through the grass, retracting their heads and making hissing noises if we got too near. These creatures were the only ones out of hundreds of varieties that gave any impression that they even noticed us.

Twenty years ago, the Florida Everglades also presented an ecological paradise that teemed and undulated with life, but real estate development and man-made canals have whittled away the natural ecosphere, and urban runoff and pollution are destroying the rest.

Once upon a time, the whole North American continent abounded with millions of animals and birds, as did Europe, the Middle East, Australia, and all other regions. Not only are we losing species, but we are killing millions of genetically unique individuals within single species. As someone recently said, cutting down a forest is equivalent to shooting the animals and birds that live within.

I refuse to contemplate a world devoid of diverse life forms. Is our development so important and sacrosanct that we must destroy all other species in our drive toward domination of the planet? Such behavior is anthropocentric. Let us instead develop a sense of humility and a deep love for our fellow creatures, recognizing that their value is equal to our own.

6

Overpopulation

During the nineteenth century, some well-meaning people introduced several English rabbits into Australia. Rabbits are normally considered nice, gentle creatures, but these domestic rabbits escaped and, experiencing no natural predators and liking the environmental conditions, began to reproduce with a fury. By the 1950s, the land in many parts of Australia undulated with rabbits. They ate the grass that was grown to feed sheep and cattle and provoked erosion, they ringbarked trees, and they consumed small trees and shrubs. We were at a loss to know what to do with this affliction, until scientists at the Australian Commonwealth Scientific Institute and Research Organisation discovered a flu-like virus called myxomatosis, which was highly contagious and lethal within the rabbit population. This organism was then released into the environment, and the plague of rabbits disappeared within a couple of years. Luckily, the virus remained contained and had no other side effects. We still see some wild rabbits in the bush, but they are now relatively well controlled.

Having in the preceding chapter outlined the possible plight of most other species on earth, I will now concentrate on the human species. While others diminish in numbers, we proliferate by the millions each day, like the rabbits. We, like them, have no natural predators. Having tamed the natural environment, we now rule the roost. Furthermore, the Bible says we were given dominion over the earth, and we believe it. To reiterate, in 1900 we numbered 1 billion, in 1950, 2 billion, and in 1990, 5.2 billion; by the second half of next century, we could number 14 billion. Few of us really understand the havoc we are wreaking on perhaps the only life system in the universe, and few of us seem to care. As long as we have access to our creature comforts and can buy and sell what we need, why think about the future? It's too scary. Yet beneath the veneer of comfort, we really know what is happening.

I have always lived in the Western world, thirty-eight years in Australia and fourteen years in the United States. I had never visited India, because I was too frightened to come face to face with the poverty and suffering and feel absolutely impotent—to know that there was nothing I could do to help these people. But in 1988 I was invited by a medical and scientific organization to lecture in thirteen Indian cities on the medical effects of nuclear power, and I decided to accept.

Although I had read about the Third World for years, no amount of intellectual understanding prepared me for the actual experience. Sights, sounds, smells, disease, and poverty engulfed me. Children dragging their torsos about on skateboards, using their arms to propel polio-paralyzed legs; lepers without fingers and noses begging on every corner; dark-eyed children with blond hair made light from malnutrition, bandy-legged children with rickety bent bones secondary to vitamin D deficiency; and severe cases of kwashiorkor secondary to protein malnutrition. The streets were like a pathology museum—never had I seen so much preventable disease. Drinking water came from streams polluted with animal and human sewage. Hygiene in our sense was nonexistent. Yet the smell of exotic Indian food mixed with the odors of disease and decay, and the population was vibrant and independent. Sitting on a street corner, one could see a sight a minute, one never again to be repeated. The most amazing collection of cultures, costumes, religions, caste systems, and peoples gathered together, living and working in relative harmony.

As I traveled the country, I was distraught to discover that the money of the very rich people in India would suffice to house, feed, clothe, and heal most of the people in their country. These people, born and bred to wealth, diamonds and gold adorning their bodies, walk among the poor and dying with unseeing eyes, much as rich New Yorkers step over the bodies of alcoholics and schizophrenics lying and dying in the gutters of New York City.

In Bombay, a large proportion of the population dwells in tents and lean-tos constructed from scraps of garbage, perched on top of the massive rubbish dumps that circle the city. They make their living by scavenging any piece of rubbish that can be retrieved and sold. The air is pungent with the smell of rotting debris and, on a hot, humid day, almost suffocating. Yet from these hovels emerge the most elegant, dignified women, draped in colorful saris—thin and fat, old and young, all with a grace and beauty that defies description. Each family often owns only one sari, so the other women stay at home that day. Even in the face of great deprivation and poverty, human dignity manages to prevail.

It was with some difficulty that I tolerated this physical assault upon my refined Western senses, until at the end of two weeks I became mortified and angry. I ranted at the poverty and disease, at the seeming heartlessness of the rich, and at my sense of despair and impotence. I hated India and wanted to escape. But the feelings of desperation passed within twenty-four hours, and I became more comfortable. As the morbid conditions grew less obvious, I was able to look at the scene with a sense of detachment and a degree of psychic numbing, or denial, that gave me a modicum of comfort. I suppose that is how the human psyche adjusts to difficult situations.

In the south of India, where coconut palms and banana trees hang over beautiful canals and the air is warm and moist, a population explosion is in progress. The Muslims are attempting to outbreed the Hindus in a competitive, almost warlike fashion. I heard of one Muslim man who boasted having five wives and sixty children. I had never seen such a plague of children, and it was obvious that if drought or famine struck this population, a large number would perish. On a small planet with finite resources, such reproductive behavior is altogether inappropriate. India currently has a total population approaching one billion people. Organized birth control was attempted in the 1970s by Prime Minister Indira Gandhi and her government. Hundreds of thousands of men were sterilized by vasectomy, having succumbed to the incentives of free transistor radios and other goodies. But eventually the people rebelled against this coercion, and birth control is now again a big problem in India.

In regard to population control, China, by contrast, offers a good example to the rest of the world. Its population now exceeds one billion, but it has been kept in check over the last several decades by a law mandating that each family have only one child. Tax incentives were offered and financial fines imposed if people did not comply. The results of this policy are quite fascinating. A walk down any Chinese street at dusk is like strolling into a scene from the movie *The Last Emperor*. A three-year-old child struts along the footpath, followed by two adoring parents and four adoring grandparents. What effect this "only child" syndrome will have upon the child's psychological development is at this point a matter of speculation.

Unfortunately, Chinese society still prefers boy to girl babies, and female infanticide is not uncommon. Aside from the feelings of repugnance it evokes, this cruel practice reflects an ignorance that is anachronistic. In terms of global survival and from a strictly biological perspective, female babies are definitely more valuable than males, because a woman produces only one egg per month and it takes nine months for gestation to occur, whereas a man produces millions of sperm per ejaculation and so in theory could father hundreds of children. This does not, of course, mean that I am advocating infanticide.

Certain male-dominated religions have dogmas related to

human reproduction that are also obsolete. These religious organizations play a huge role in the progressive overpopulation of the world. Procreation and sexual rituals have for centuries figured prominently in the Hindu tradition, and polygamy is integral to the Muslim faith. The Roman Catholic church is run by celibate men who have never experienced sex, or at least never should have. Yet Jesus did not lecture his followers on the subject of birth control, contraception, or abortion. Indeed, abortion became part of the Catholic teachings only some one hundred years ago. In a world threatened with extinction because of overpopulation, the pope continues to exhort people to have more babies. In Dublin, Ireland, I saw a scene that I think illustrates contemporary Catholic tradition: a young, unkempt, harassed woman staggering along in high heels, carrying a screaming toddler, followed by a husband pushing a pram containing a tiny infant, and in their wake, six other young children of various sizes and ages. Abortion and divorce are illegal in Ireland. However, contraception was progressively legalized in 1979 and 1984, so that now it is totally legal. Moreover, a woman, Mary Robinson, who believes in birth control, abortion, and divorce, was recently elected president of Ireland, and at last there is now some hope for enlightened political change with regard to these important issues.

It is clear that the reproductive policies of certain religions are out of date if life on earth is to continue. I am not suggesting that the personal quest for spirituality and enlightenment, as exemplified by the teachings of Jesus, Buddha, Allah, or the Hindu gods, is obsolete. But it is beyond time for women to assume leading roles in the earth's representative religions and to establish new and more appropriate spiritual organizations. I do not make this statement lightly, but with a sense of gravity and urgency.

Just under 50 percent of the human population is composed of women. According to the United Nations, women do two-thirds of the world's work, including much of the work of producing food, and for this labor they receive 10 percent of the global income and own only 1 percent of the property.[1] And they have very little power. Because women give birth to all the babies, they bear a major responsibility for the population explosion and are thus in an extraordinary position to stop the exponential increase of *Homo sapiens*. (Exponential growth is a mathematical concept not difficult to understand. For example, a water lily in the middle of a huge pond takes four weeks to reproduce. It then takes four weeks for another doubling, from two to four lily leaves. After ten years of steady growth, the pond is half full, but the next doubling of four weeks duration will fill the pond.)

It is a well-known fact that in places where women are well educated and relatively affluent, the birthrate is low. Because they have access to contraception, they understand the reproductive cycle and can therefore control their own destinies.

Poor women tend to have many babies, because they are never sure whether their children will survive the diseases and malnutrition of infancy. Furthermore, poor parents need adult children in their old age as a form of insurance—to grow food, to run the farm, and to care for them. Forty thousand children die each day from malnutrition and preventable disease like diarrhea, measles, tetanus, whooping cough, polio, and diphtheria. Eighty percent of all disease in the poor developing world is caused by a lack of access to clean water.[2] (In fact, community health is measured better by the number of water taps in a village than by the number of hospital beds.) Some 1.7 billion people do not have access to clean water, and 1.2 billion do not have adequate sanitation. Malnutrition predisposes infants and children to contagious diarrhea and other infectious diseases. Hence the very high death rates among children in the Third World.[3]

Contraceptives of all varieties must be made available to the reproductive men and women of the world. Every person has as much right to birth control as to food. According to the United Nations, over half the 463 million married women in developing countries, excluding China, need contraceptives because they do not want more children. Yet most of these women are poor and uneducated and do not understand the ovulation cycle, let alone how to use contraceptives safely even if they had them.

Vasectomy is not popular among men, although a recent one-year study by the World Health Organization found that weekly injections of a synthetic male hormone "can maintain safe, stable, effective, and reversible, contraception."[4] There are some contraceptives very appropriate for Third World populations. For example, injectable Depo-Provera is a suitable hormonal suppressor of ovulation, as is a preparation called Norplant, which can be implanted under the skin and which remains effective for five years. A new quick and easy contraceptive is a drug called RU 486. It is a hormone developed in France that, if taken within days or weeks after conception, induces a painless, spontaneous abortion. Years of testing have revealed virtually no serious side effects.[5] Yet the U.S. Food and Drug Administration refuses to condone its use, because of archaic political and religious objections. If I could, I would like to be a world salesperson for RU 486—what a difference this easy form of birth control could make to overpopulation!

The debate about abortion in the United States strikes me as obsolete and misogynist. Thirty million species are endangered by our relentless procreation, yet we still argue about abortion as if we lived two centuries ago, when the human population stood at one billion. So invidious is this absurd American debate that during the Reagan years funding for overseas birth control aid was cut off to countries that permit abortion, including, of course, almost all of the developing world. As a result, U.S. population assistance programs fell 20 percent between 1985 and 1987, from $280 million to $230 million.[6]

Compare these small sums with global military spending. President Reagan spent more money than all past presidents combined, and most of it went to the military-industrial complex for high-tech weapons systems, many of which were used in the Persian Gulf war. The world spends a total of $1 trillion per year on weapons, and the World Bank estimates that only $8 billion a year needs to be spent to provide global birth control.[7] But according to the World Watch Institute, another $25 billion a year must also go to social improvements in the Third World to reinforce birth control policies, so as to stabilize the world population at eight billion by 2050. Improved socioeconomic status leads to better prenatal care, which leads to lower maternal and infant mortality. Third World maternal deaths, which number half a million a year, are related to malnutrition, prenatal anemia, and concurrent infectious disease.[8]

Almost half the abortions in the world are illegal, and women often die dreadful deaths from gram-negative septicemia, or blood poisoning, resulting from dirty instruments. Deaths are rarely associated with legal abortions, but in countries where abortion is illegal, rich women *always* get their safe abortions, while poor women often die. Years ago when I worked in the casualty department of the Royal Adelaide Hospital, a woman was brought in on a stretcher, desperately ill from a septic illegal abortion. She was accompanied by a weeping husband and six terrified children. She was going to die and would leave her family behind her—all because abortions were not legal at that time and because she could not cope with more children.

In my years in general practice, it was often the Catholic women who knocked on my door when their thirteen-year-old daughters were pregnant, or when they themselves became pregnant by a man who was not their husband. It was clear that religious principles often went by the board when the practical issues of life and death presented themselves.

Urgent global educational efforts must be conducted in regard to sexually transmitted disease. AIDS is spreading like wildfire in Africa and elsewhere. Sexually spread wart virus is related to cervical cancer, vulval and vaginal herpes can cause herpes encephalitis in a newborn baby, some forms of syphilis are drug resistant, and gonorrhea is a nasty disease.

It may seem somewhat contradictory for me to stress, in discussing overpopulation, that AIDS and other diseases must be prevented. Yet the cure for overpopulation is not epidemics of disease, or nuclear war, as some people suggest. It is redistribution of wealth, compassionate politics, and caring societies. Logically, if women and men are well fed, well educated, and financially secure, their children will not die in infancy and the birthrate will automatically drop.

I would like to cite Cuba as an example of a country with good medical care and equitable distribution of wealth. Before the U.S.–backed dictator Fulgencio Batista was overthrown by Fidel Castro, Cuba was largely controlled by American business interests, the Mafia, and wealthy American tourists.

When I visited Cuba in 1979, I learned that before the revolution the majority of the population was severely repressed, illiteracy was endemic, and more than 70 percent of the people suffered from anemia, hookworm, intestinal parasites, and tuberculosis.[9] The introduction of communism certainly brought with it political repression and infringements of civil liberties, but the population at large was liberated. Several years after the revolution, almost all adults were literate, all children attended school, prenatal clinics were established, and prenatal care was mandatory for all women. Infant and maternal mortality rates plummeted, ranking among the lowest in the world. Free medical care became available to all, and as the standard of living increased, the birthrate fell.

My visit to Cuba was a wonderful experience, but I felt distressed that the United States had ever since the revolution applied a trade embargo on sugar exports and on such imports as medical drugs, machinery, oil, and other necessary items. Despite these restrictions, Cuba had demonstrated that a small, isolated, depressed country could raise the standard of living for its people.

We were free to wander at will through city streets and country roads, and we talked to many people. One old peasant woman broke into a huge grin when I asked her what she thought of the revolution and of Castro. She said with enormous pride, "My son's a doctor." He would never have received that educational opportunity while the country was under Batista's rule.

It is clear that the United States was not interested in allowing Cuba to become a model for other Central or South American countries, lest U.S. corporations lose control of cheap cash crops, cheap labor markets, and access to cheap natural resources. In general, the American government is not concerned about the condition of the repressed indigenous populations in many of these Latin American countries, because the repression actually promoted the productivity of multinationals. This is why the CIA organized and financed a covert Contra war against the Nicaraguan people when they attempted to emulate the Cuban policies of excellent education, health care, and social welfare after they ousted the American-appointed dictator Anastasio Somoza. Within months of its revolution, Nicaragua was able to decrease infant mortality rates and the incidence of malaria and to improve literacy by 30 to 50 percent.[10] But the CIA-backed Contra forces prevailed, and by 1989 the back of this wonderful social revolution was finally broken. Now many Latin American countries are locked in social situations similar to prerevolutionary Cuba's—shocking poverty, generalized malnutrition, illiteracy, disease, and rampant overpopulation.

Bearing in mind the conditions of the Third World, we must be acutely aware that the problem of overpopulation will eventually also become critical in parts of the developed world that are still reproducing beyond a stable population base. For instance, the U.S. population, which stood at 249 million in 1990, is growing at about 2.2 million per year and is expected to reach 268 million by the year 2000. It is growing faster than that of eighteen other industrialized nations.[11]

All of us must therefore accept responsibility for our reproductive habits. No longer can we say that it is God's gift that we are pregnant or that it is okay for us to have babies because we are wealthy and our population growth rate is slower than that of many Third World countries, while we exhort them to slow down. Those attitudes will be seen as racist and elitist. We must all take responsibility and realize that we are suffocating thirty million other precious species by our egocentric and anthropocentric attitudes. It is time for the global community of *Homo sapiens* to take the radical step of limiting families to one child. This policy could be formulated by the United Nations. The nations of the world could thereby assist and aid one another while taking responsibility for their own societies. Of course, birth control must be voluntary on the part of educated individuals, endorsed and supported by government subsidies and help, rather than mandated by law. Such behavior could become fashionable and an integral part of the "new world order" after appropriate national and international educational regimes.

Women are in all ways crucial to planetary survival. In the developing countries, they do most of the agricultural labor, they understand the rotation and planting of specific crops, and they accept the responsibility to provide good food for their families if they can. For instance, in sub-Saharan Africa, women produce four-fifths of the food. Yet it is the men who own all the tools, production facilities, roads, trucks, and land and who make all the key political decisions. Because the contribution of women is almost totally unrecognized, they are officially said to make up 20 percent of the agricultural work force.[12] In Nepal and India, it is the women who slave in the fields and who carry huge sacks of rice slung from Hessian straps around their heads, while the men tend to sit around drinking tea and coffee, philosophizing about life. Similar scenes are frequently seen in TV shows that portray life in African countries.

Several stories told in *The Global Ecological Handbook* illustrate the hazards involved when women's work is ignored.[13] World Bank reforestation projects in Kenya and India failed when the education and advice were directed toward men, but they succeeded when women were taught to care for the trees. In Gambia, development projects were initially removed from women's control, but when women were given land rights and provided with seed, equipment, and centers for child care, rice production increased sixfold.

We ignore women's contribution to world politics at our peril. It is time for men in organized religions, in politics, in corporations, and in global agriculture to stand aside and attest to the intelligence and wisdom of women. The birthrate will fall, wealth will be redistributed, compassionate societies will evolve, wars will almost certainly cease, and food will be efficiently produced and distributed.

According to the United Nations report called *Our Common Future,* women are the single most important sector of society when it comes to lowering population growth and caring for the earth.

7

First World Greed and Third World Debt

When I think of malnutrition and of forty thousand children around the world dying daily from starvation-related disease, I am brought brutally face to face with the affluent countries.[1] The gap between the 20 percent rich and the 80 percent poor is not decreasing but increasing. In 1987, the average income per capita in the First World was $12,070, while it was *$670,* or 6 percent of $12,070, in the Third World. Ten years earlier, this number was 9 percent, so the income of the poor actually decreased.[2]

Access to food is a preoccupation of 80 percent of the world's people, yet food in the United States is overabundant, and many Americans consider it a difficult substance to understand and sometimes even poisonous. It is true that fruit, vegetables, and processed goods are often contaminated with toxins, pesticides, artificial dyes, hormones, and antibiotics. On the other hand, life expectancy continues to rise in the United States, to over seventy years of age, because of a combination of preventive medicine, medical care, and adequate food, while in the Third World it progressively declines, to the midthirties in many countries. In

India, I saw only two people with gray hair. The people of the First World—Americans especially—are obsessed with vitamins, cholesterol, sugar, and the like, while the majority of the population battle obesity or, to put it simply, overnutrition. In 1988, it was estimated that 950 million Third World people did not consume enough calories for a normal active working life.[3]

I am repulsed when I go out to dinner and read on the menu, "All you can eat for $10," and am served a steak weighing one pound and a huge potato replete with sour cream, a salad with a choice of ten salad dressings, and bread soaked in garlic butter. I know that more than half the food will be thrown away, because my stomach's capacity is too small and I do not need all that food, while forty thousand children die daily of starvation. Fifteen percent of the food used by U.S. homes and restaurants is thrown away, and this food is worth $50 billion per year.[4]

It is no wonder that if you have too much food, you become obsessed with it. In fact, people in wealthy countries need to eat less, to eat items lower on the food chain, and to share their riches. The grossly excessive amounts of food produced in the United States should be exported to Third World countries that really need it. But conditions must be imposed on the exporting activities so that Third World farmers are not jeopardized by the dumping of cheap imported food on their markets.

Apart from the poor people who live on the fringes of U.S. society, most Americans probably have ample or excessive amounts of vitamins, proteins, carbohydrates, and fats in their diet. Although we must be aware that in the wealthiest country on earth in 1985, twelve million children and eight million adults were malnourished because of inadequate or inappropriate food intake.[5] But from a global perspective it is immoral that a small minority of people in the world are overnourished, while most are undernourished. We should also understand that severe malnutrition in childhood induces mental retardation, because the developing brain needs to be well fed. This means that millions of people are unable to better their condition, because their mental capacity has been damaged.

Studies reveal that Americans eat twice as much protein a day as they need, and it is clear that many common diseases in the U.S. population are related to the problem of overnutrition.[6] For instance, excessive or daily consumption of red meat increases the incidence of colon cancer in women[7] and probably in men; it also induces high cholesterol with associated hypertension, atherosclerosis, heart attacks and strokes. Moreover, the obesity caused by overeating can lead to osteoarthritis and joint conditions. Hypervitaminosis disease can also result from excessive doses of vitamins A and D, and many Americans take large quantities of extra vitamins, when, in fact, they already obtain their daily requirement of vitamins by eating a normal, balanced diet.

What can Americans do to make sure that they are nourished better and that their eating habits can help benefit the global population?

- Eat foods lower on the food chain. A high intake of fruit and vegetables ensures both a high-fiber diet, which decreases the incidence of colon cancer, and a low-cholesterol diet.
- East less meat. Most red meat and some chickens are contaminated with antibiotics, the same drugs doctors prescribe to kill bacteria in infected patients. Widespread, indiscriminate antibiotic consumption by the general population leads to drug-resistant bacteria, precluding important therapeutic approaches to serious human bacterial infections. Steroidal hormones are also fed to cattle, in order to increase the weight and muscle content of the meat, thus maximizing the farmer's profit.
- Eat more grain. In the early years of this century, Americans obtained 40 percent of their protein from grains and cereals, but now only 20 percent comes from vegetable products. Meanwhile, their consumption of red meat doubled between 1950 and 1975.[8] Hundreds of years ago, when our forebears lived as tribal nomads, the diet consisted mainly of fruit, vegetables, and grains. But with the advent of more "sophisticated" life-styles, the consumption of animals increased. In biological terms, our bodies did not evolve to

depend on large quantities of animal protein—it is not good for us!

Furthermore, just as the world environment cannot sustain 5.2 billion people, it can hardly sustain the present global population of four billion cows, sheep, pigs, and goats and nine billion fowl. It takes sixteen pounds of grain and soybeans to produce one pound of beef, six pounds for one pound of pork, and three to four pounds for one pound of poultry or eggs. Animals now eat ten times more grain than do the American people, enough to feed the entire U.S. population five times over. The production of each acre of food uses energy equivalent to 150 gallons of oil, and the processing and packaging of food alone accounts for 6 percent of the U.S. energy consumption. From 145 million tons of grain fed to animals, only 21 million tons of meat, chicken, and eggs are produced.[9] What an extraordinary waste of soil, water, fertilizers, pesticides, and energy—and all for what return but increased heart attacks, obesity, and strokes?

The Feinstein World Hunger program at Brown University has estimated that if the world population is fed primarily with grains and vegetables, there is at present enough food to ensure the United Nations' recommended daily per capita calorie intake of 2,350 calories for six billion people.[10] While billions starve, one-third of the grain grown in the world and half the fish caught are fed to animals in rich countries. If the world's population reaches eleven billion, its annual food production must increase two and a half times merely to maintain the current situation and low per capita output.[11] We have work to do!

Sixty percent of the world's best agricultural land is situated in only twenty-nine countries, which house 15 percent of its population. A mere 11 percent of the land area of the planet is suitable for agriculture, yet soil erosion associated with deforestation, dam construction, and irrigation is depleting 77 billion tons of the world's topsoil each year. Our global treasure is being lost into the sea and silting up rivers. In addition, one million hectares of rich farmland in the United States are destroyed by "development"—covered with asphalt, freeways, houses, and shopping centers—equal to an area double the size of the state of Delaware. The use of energy to grow crops in the United States has increased 300 percent since 1945.[12]

In a world where equitable redistribution of wealth is vital, let us now discuss the financial support that the wealthy 20 percent currently give to the poor 80 percent. The United States will be our example. In 1987, it allocated $13 billion, or only 0.19 percent of its gross national product, for foreign aid, but $4.8 billion of it went for military equipment. Only $1.5 billion was for food, $3.2 billion for economic security, and $2.5 billion for development assistance. Furthermore, over half the economic and security aid was disbursed to Egypt and Israel.[13] (In 1980, U.S. aid to Israel per capita was 120 times the U.S. aid to India.) In other words, the most needy countries were shortchanged.[14]

Most aid serves as an instrument of foreign policy, not really as a charitable gift. For example, in 1965–66, during a famine, the United States threatened to cut off food aid to India when its government attempted to take control of U.S.-owned fertilizer companies. India capitulated because it needed the money, thereby giving more freedom to U.S. investment companies. In effect, while millions of Indians starved, food shipments were stalled to force the government to capitulate to the demands of U.S. corporations. In 1964, U.S. aid to Brazil dropped from $81.8 million to $15.1 million because America disapproved of the government at the time.[15] These are just two instances in which the U.S. government withheld food for political purposes. Food is used to reward and manipulate poor countries rather than to feed hungry people.

Surprisingly, most U.S. aid actually winds up subsidizing American corporations. During the Johnson administration, 90 percent of all foreign aid benefited U.S. corporate development programs, such as the building of dams, nuclear power plants, roads, and bridges in the Third World, and the profits accrued to the relevant U.S. companies. So U.S. foreign aid serves not only as a coercive instrument of foreign policy but also to support private U.S. contractors, universities, banks, consulting firms, lobbyists, and so forth. In fact, foreign aid is now recognized to be a lucrative business, and companies are scrambling to capitalize on it. Even in 1970, multinationals invested $270 million in Africa and repatriated $995 million, $200 million in Asia and received $2,400 million, and $900 million in Latin America for $2,900 million. Corporations also tend to borrow most of their investment funds for Third World projects from Third World banks.[16]

In 1986, U.S. foreign aid expenditure totaled $15.9 billion, while Americans spent $10.3 billion on movies and theaters, $34.2 billion on tobacco, and $59 billion on alcohol.[17] An expenditure of five cents per person would save the sight of 100,-000 children who are blinded annually because of a vitamin A deficiency, and a mere three dollars each would immunize them against poliomyelitis, tetanus, whooping cough, diphtheria, and measles. One year's expenditure by the U.S. cosmetics industry would provide 1.6 billion people with sanitation.[18]

In percentage of GNP given away as foreign aid, Norway tops all countries, with 1.1 percent. The Netherlands spends 0.98 percent. The United States ranks next to last out of eighteen countries, with, as we have seen, 0.19 percent of its GNP.[19] Even the best of these figures is still very small.

How does the United States spend its money? In 1987, America ranked first in military spending, arms exports, military technology, naval fleets, global military bases, and number of nuclear bombs; fifth in literacy rate; seventh in per capita spending on public education; eighth in life expectancy; sixteenth in the number of women enrolled in university, eighteenth in infant mortality, twentieth in school age population per teacher, and twentieth in population per physician.[20] These data, though not strictly financial, indicate present priorities in American society.

The United States spends about $300 billion per year on weapons, personnel, and military equipment, while the total global expenditure on them is $1 trillion. One-third of the money spent on one Trident submarine, $70 million, could eliminate malaria in the world by clearing mosquito-ridden waterways and providing prophylactic medicines to the target population, and one week of the global military spending of $20 billion could put an end to all starvation.[21]

To quote Thomas Jefferson, "The care of human life and happiness, and not their destruction, is the first and only legitimate object of good government." Or as Dwight D. Eisenhower put it, "Every gun that is made, every warship launched, every rocket fired, signifies in the final sense a theft from those who hunger and are not fed, those who are cold and not clothed."[22]

International aid is but a Band-Aid on the wounds of Third World suffering. The people there are not just malnourished and deprived because of overpopulation, inadequate distribution of money, lack of education, or bad land management. They are poor and starving because financial powers in the developed world exploit them to satisfy their own greed and continued affluence.

Two major dynamics occurring globally are having, and will have, severe and lasting repercussions upon the developing world. One is the huge foreign debt incurred during the oil crisis in 1973–74; the other is the dangerous set of negotiations taking place in the organization called the General Agreement on Tariffs and Trade (GATT).

Now, I recognize that some of this information can appear boring and dry, but these economic realities will shape the destinies of most of the world's people and thirty million species. We are therefore obliged to study hard, to think, and to act.

THIRD WORLD DEBT

Third World debt is exacerbating global environmental degradation. Until 1973, the poor developing countries had made great strides in public health and preventive medicine, in education, and in crop production. But in the years 1973–74 world dynamics changed. The oil-producing, OPEC countries suddenly increased the price of oil fivefold, and oil became scarce. There were large and often angry lines outside gas stations in the United States and elsewhere. The oil-rich countries made huge profits and deposited their money, called petrodollars, in the world's major banks, particularly in the United States.[23]

The banks, ever eager to profit from this windfall, decided to offer low-interest loans to Third World countries, which could obviously benefit from extra cash. The banks called this lending policy "recycling," but it turned out to be, in effect, an international lending spree, in which they unloaded petrodollars on unsuspecting developing countries.

The World Bank, which represents many of the major banks in the United States and other wealthy countries, decided at a conference in the Philippines in 1976 to lend this impoverished country $4.5 billion. It openly stated that it would rather lend to "stable" countries (that is, ones ruled by dictators) than to democracies governed by the people. Hence President Marcos of the Philippines was an excellent client. Countries with similar dictatorships received huge petrodollar loans—Chile, Brazil, Argentina, and Uruguay. Other borrowers included Mexico, Tanzania, and many struggling African nations. The amount of money available was vast. Lending to Latin America increased from $35 billion in 1973 to $350 billion in 1983.[24]

Most of the loans were based on variable interest rates. During 1981–82, U.S. interest rates doubled, to 20 percent, because of tight monetary policies caused by the growing U.S. deficit, engendered by the Reagan administration's lavish spending on weapons. But each 1 percent increase in the interest rate meant $4 billion more that Latin America had to give American banks in interest payments. Furthermore, a global recession at that time decreased the demand for exports from developing countries, thus decreasing their income. Debt servicing rose from 15 percent of their export earnings in 1980 to 31 percent in 1986 in sub-Saharan Africa. Between 1973 and 1980, Third World debt increased by a factor of four, to $650 billion, and it now stands at $1.3 trillion. This is an unbelievable burden for countries whose populations are hardly surviving.[25]

To make this situation more graphic, imagine that you accepted a variable rate mortgage (vrm) of 8 percent in 1981 for $100,000 to buy a house. But instead of being "capped," as most vrm's are, at some arbitrary level—say, 15 percent, yours had no legal ceiling. When interest rates went up, therefore, so did the interest on your mortgage—from 8 percent to 20 percent! But at the same time your income failed to increase, or even decreased. Instead of $800 a month in mortgage payments, you were forced to pay $2,000 per month. You would quickly become bankrupt. This was the plight of many Third World countries.

But worse was to come. When oil prices fell in the early 1980s, a crisis developed in Mexico, an oil-exporting country. Nobody quite grasped how severe the situation had become until 1982, when the Mexican finance minister visited Washington and announced that his country could no longer continue payments on the $80 billion debt. Suddenly, the U.S. government and banks realized that the global financial system was on the brink of collapse. The banks had lent more than 100 percent of their capital to Third World countries, which appears, in retrospect, an extraordinarily irresponsible policy for them to have pursued.[26]

A state of emergency existed, and heads of banks, officials from the White House, and the U.S. Treasury met for two weeks behind closed doors to try to deal with it. The World Bank and the International Monetary Fund (IMF), which represents most of the major banks in the Western world, produced a proposal that basically demanded that developing countries sacrifice government spending on health, education, and welfare in order to service the debt and that they increase the export of commodity or luxury goods to earn more money. The IMF became crisis manager number one and instructed banks how to contribute to the Mexican rescue. Mexico had to sign a letter of intent to earn more and spend less.[27]

Before I turn to the dire human consequences of these restrictive policies, let me describe how parts of the loans had been misappropriated by people in the developing countries. Nearly half the money lent to Mexico was not used to benefit the people but was instead invested by rich Mexicans in foreign banks and for the purchase of foreign luxury apartments (in other words, the loans were stolen from the Mexican people).

Between 1980 and 1982, $40 billion ended up in America. And during this time U.S. banks actually went to Mexico to solicit money derived from U.S. loans by rich Mexicans to be deposited back into U.S. banks. In short, there was no overall stipulation about who controlled the loan money or how it was to be spent.[28]

In Latin America, for instance, a large percentage of the petrodollars was used to buy expensive weapons. Military spending in these countries increased 10 percent each year during the 1970s and 1980s. In Africa, spending on weapons increased 18 percent annually. The Stockholm International Peace Research Institute estimates that 20 percent of the Third World debt is a result of military spending.[29] It goes without saying that these poor countries do not manufacture military equipment themselves. It is made in the United States, the Soviet Union, Israel, Germany, France, Britain, and other industrial nations. Each country's military-industrial complex lobbies to sell weapons to poor countries. Arms bazaars are held in many cities of the United States and other countries each year. There, phallic missiles and planes are displayed, often draped with almost naked women in bikinis. Sheiks, Third World potentates, and military men amble around the displays, idly deciding which nice missile they will purchase that day.

President Marcos misappropriated millions of dollars lent to the Philippines and deposited them in Swiss and American banks, while millions of his people were severely malnourished and lived in absolute poverty. The banks seemed to support and indeed condone such behavior. The money also wound up in other inappropriate ventures. For instance, foreign loans were used to finance a nuclear power plant constructed in the Philippines on an earthquake fault, near two active volcanoes. The reactor—never operational, because of the obvious danger—was built by Westinghouse. This fiasco now costs the Philippine people $500,000 per day in interest alone, and the profit naturally went to Westinghouse. Unfortunately, when President Corazon Aquino was democratically elected, she was forced to put debt payment above priorities for health, education, or housing, thereby violating the Philippine constitution. In 1989, about 44 percent of the Philippine national budget went to repay foreign debt, and most of the people continue to live in dire poverty.[30]

Loans to other countries were used to finance badly planned highway systems in big cities inhabited by the rich, as well as railroads, huge dams, and power plants. These projects had no relevance to the millions of people trying to sustain an existence with poor land, archaic equipment, minimal housing, and few, if any, health care facilities. And all profits accrued to the foreign corporations that carried out the construction.[31]

The World Bank has encouraged countries to destroy tropical rain forests to pay back their debt. In Brazil, it proposed that 5 million hectares of the Amazon be "brought under control and management"; in Ecuador, 1 million hectares; and in India, 30 million hectares. It is the same story in the Congo and in Papua New Guinea. The World Bank also approved a loan of $156 million for a dam on the Serang River, in central Java, Indonesia, which would entail the resettlement of twenty thousand people and the flooding of huge areas of forest. Similar projects have been prepared in Zaire and in Gujarat, India.[32]

Since the Mexican crisis in 1982, the IMF has decided to impose severe austerity measures that forced desperately poor Third World countries to use more and more of their land to grow cash or luxury crops for export, such as bananas, coffee, cocoa, pineapples, and flowers, in order to help pay off the debt. Forests and virgin land are being destroyed and cultivated, producing ecological damage and leaving very little land on which the indigenous population can grow food. The people then become dependent upon cheap imported food from the United States (grown with subsidies from the American government). In the 1970s, food imports to Africa increased 600 percent, and by 1985 two-fifths of Africa's food was imported. The net effect of these policies has been to produce nations that are hostage to the American and European agricultural system and that lose their independence. Instead, foreign aid should be used primarily to establish agricultural and economic autonomy for Third World countries.[33]

Women tend to bear the brunt of these IMF policies, for they spend more and more of their day digging in the fields by hand to increase the production of luxury crops, with no machinery or modern equipment. It becomes their lot to help reduce the foreign debt, even though they never benefited from the loans in the first place. Their health suffers, and they become tired and anemic because of poor diets. Consequently, maternal and infant mortality is rising in many of these countries.[34] The price of food, too, is increasing, and wage cuts imposed by the IMF are severe. Millions of people can no longer afford adequately to feed, clothe, or educate their children in the Third World. One-fifth of the agricultural products grown there are exported to the First World.[35] Meanwhile, as world markets become flooded with these commodity or luxury exports, the prices drop and the First World benefits from cheap luxury imports (from 1984 to 1986, the United States saved $65 million on raw materials from the Third World because of price declines). In 1986, these price cuts reduced the income of sub-Saharan Africa by $19 billion—nearly four times the amount promised in emergency aid that year.[36] The First World determines the price of Third World–produced luxury goods. The Third World has no say![37]

Most of the profits from commodity sales in the Third World go to retailers, middlemen, and shareholders in the First World. Only 15 percent of the $200 billion in the annual sales of these commodities in rich countries winds up in the countries that grow them. For instance, just 11 percent from the sale of bananas is paid to the producing country. In Brazil, only 4 percent in royalties is paid for the millions of tons of bauxite (used to produce aluminum) exported to the United States. The workers in the Third World in general receive only 1 to 2 percent of the value they create. Nicaraguan coffee pickers receive 0.5 percent of the wholesale value of the coffee, while governments and rich plantation owners make $750 million per year. And tea pickers in India starve because they get only eight cents per day and food prices are high, whereas company profits in 1976–77 were $21 million. In the Caribbean, the people actually starve beside fields growing flowers and tomatoes for export to the United States.[38]

Wealthy countries impose tariffs or trade barriers on processed goods, but none on raw materials, thus ensuring that poor countries remain in poverty. For instance, in 1985, British tariffs on raw cotton were zero, on cotton yarn 8 percent, and on cotton T-shirts 17 percent.[39] So the Third World can never break the poverty cycle, because First World tariffs work against the importation of manufactured goods from the Third World. A Third World country is defined as one that exports raw materials and imports finished goods. But processed goods are worth much more money than raw materials are.

And so the spiral continues: increased debt leads to more cash crops and environmental degradation, which leads to flooded markets in the First World and lower prices, with decreased return to the Third World. Therefore, the debt increases, and this leads to malnutrition, starvation, and helplessness.

To make matters worse, the IMF also decreed that governments of debtor nations severely curtail spending on social programs. In Mexico, whose economy has stagnated for seven years, there has been a severe overall decline in health services, education, and housing. Twelve million people are out of work; the Mexican people consume one-third fewer calories than they did in 1982, because food is more expensive. Two million acres of arable land were taken out of public agriculture because of an 80 percent cut in public funding. About 40 percent of the money that Mexico earns from exports goes to service the debt, and real wages for its people have fallen 30 to 40 percent. Mexican interest payments are $1 billion per month, the debt has actually increased by 50 percent, and the country now owes more than it did before the crisis began. And the U.S. banks call Mexico a success story—for themselves![40]

Huge U.S.-based corporations called agribusinesses also grow food in Third World countries for foreign markets, mainly the American, and the debt crisis ensures them a plentiful supply of cheap, if not slave, labor and cheap land. They pay virtually no taxes to the host country. Ninety percent of the protein fed to British animals comes from underdeveloped countries. The meat consumption of people in the First World accounts for a volume of grain that would feed 1.2 billion people.[41]

In Tanzania, which organized the best health care system in Africa in the 1970s, there has been a two-thirds cut in real wages, and the system is now in rapid decline. Hospitals are collapsing, and maternal and infant mortality is on the rise. School buildings are falling apart, and many children no longer have the access to primary education that their predecessors had in the 1970s.[42] This is a tragedy of monumental proportions. Even worse, it is typical of all the debtor nations.

By 1983, forty-two debtor countries were locked in repayments. Seven of them are responsible for nearly half the total Third World debt—Brazil, Argentina, Venezuela, Mexico, Indonesia, the Philippines, and South Korea. Some observers say that the international banking system would be threatened if any defaulted on their payment. In 1980, the net funds transferred as loans from the First to the Third World were $40 billion, but in 1983 the situation was reversed, and the debt the Third World owed to the First was $20 billion greater than the loans it received.[43] UNICEF estimates that half a million children die each year because of the debt crisis. The four Latin American countries together owe $50 billion, and theirs are disastrously poor economies. How can they possibly pay this?[44]

In Britain, the banks actually put aside or forgave half of the world debt and, in so doing, gained a huge tax relief. In fact, they made enough money from this strategy to immunize 450 million children against preventable diseases. They didn't, of course. From 1987 to 1989, using these maneuvers, British banks made £1.6 billion, which is more money than the British government paid to poor countries in foreign aid.[45] So the banks make money whichever option they choose—all on the backs of millions of starving, uneducated, sick children.

Third World debt does produce complications within industrialized nations other than the inconveniencing of the banks. Normally, poor countries buy and import manufactured goods from the First World, but instead of importing a surplus of $1.3 billion of goods, as it did in 1980, when Latin America incurred a $14.1 billion deficit in 1987, they could no longer afford to buy products made in the First World. As a consequence, the

United States lost two million manufacturing jobs. Meanwhile, ironically, the profits of the six largest commercial U.S. banks rose by 60 percent between 1982 and 1986. So the American workers suffer and the Third World suffers, but the banks flourish.[46]

Inflation in the debtor nations is horrendous. In 1989 when I visited Rio de Janeiro, I could not quite grasp that new Brazilian bank notes had been printed three times in 1989. I never did come to understand the value of the different currencies, and only later did I discover that the inflation rate in 1989 reached about 2,000 percent. I also learned that behind the façade of great wealth in Rio and Brazil, owned by 5 percent of the population, existed 86 million people who suffered from malnutrition.[47]

The total amount of money owed by the Third World is $1.3 trillion, which is a little more than the $1 trillion the global community spends on weapons each year.[48] Debt is not a way of solving the dilemma of millions of people, and death is not a way of solving conflicts. War is obviously obsolete in the nuclear age. If $1 trillion were transferred annually from the death industry to the life industry, 80 percent of the world could feed, clothe, house, and educate itself and provide excellent modern health care and contraceptive services, while the First World could concentrate on solving the problems of ozone depletion, the greenhouse effect, deforestation, global pollution, nuclear disarmament, and species extinction.

So simple—yet so difficult because of human nature. I stayed with some wealthy investment bankers recently at Palm Beach, Sydney, who told me the world would never change, because people are selfish. I suspect they were talking about themselves. It was Einstein who stated prophetically, "The splitting of the atom changed everything save man's mode of thinking, thus we drift towards unparalleled catastrophe." Man's mode of thinking—but what about women and good, caring men?

SOME ILLUSTRATIONS OF THIRD WORLD DEBT

Having described the generic problems of the Third World countries and First World banks and corporations, I now want to outline some specific conditions that require our immediate attention.

In 1984–85 when thirty-five million people were dying from drought-induced starvation in Africa, Bob Geldof and others organized immense fund-raising "food aid" concerts, which saved the lives of millions. But even so, when the drought broke, nineteen million people remained malnourished, and by 1990 another drought hit Africa. Again climatic catastrophe has combined with ongoing tribal wars, and Third World debt has caused massive population displacements and starvation. Thirty-five million people are at risk again as I write.[49]

This recent drought in Africa may be a result of the impending greenhouse effect. According to data from NASA, the British Meteorological Office, and the U.S. Commerce Department's National Oceanic and Atmospheric Administration, at Rutgers University, 1990 was the warmest year on record, and six of the seven warmest years in 140 years of recorded history have occurred since 1980.[50]

So greenhouse effect, drought, debt, and exploitation lead to misery and chaos for millions. This African drought is worse than earlier ones because although crops have failed and thousands of animals died, there is now no water in many populated areas and whole villages have been uprooted and their people rendered nomadic.[51]

In the 1960s, Africa was a net exporter of food, but over the last twenty-five years the population in sub-Saharan Africa grew at a rate of 2.8 percent per annum, while agricultural production increased at a rate of 2 percent. According to the World Bank, if present trends continue in Africa, the food gap of 15 million tons in 1990 will increase to 200 million by 2020.[52]

To sum up the world food situation, in the years 1950–84, world grain production grew faster than population, because of the introduction of new breeds of rice and wheat, which cropped two or three times a year, and because of the more intensive use of fertilizers and pesticides. Food production rose 2.6 times, increasing by 40 percent the amount available for each person on earth. In China, the per capita increase was 80 percent and in Western Europe 130 percent.[53]

Since 1984, however, grain production has decreased each year, declining 14 percent over the last four years. This decline was caused in part by the serious 1987 drought in India and by the 1988 drought in Canada, China, and the United States. Other causes for the decline were erosion of topsoil, desertification, salination, and waterlogging resulting from irrigation, depletion of groundwater supplies, and diversion of irrigation water to nonagricultural uses. In Africa, famines and drought have struck twenty-two countries. (Desertification affects the livelihood of 850 million people globally, and each year 21 million hectares, the size of Kansas, are destroyed when farming lands are converted to deserts, because of agricultural mismanagement.)[54]

The world food situation is very confusing because, on the one hand, there are millions of starving and malnourished human beings and, on the other, First World countries are producing enormous quantities of "excess" food, which is either being stored in silos and underground caves or being dumped into the sea, or dumped on Third World markets at very low prices. In 1969, the European countries, Japan, and the United States subsidized their farmers by $25 billion. At a time when the world butter price was $300 per ton, Europe paid its farmers $1,400 per ton.[55] More recently, in the United States alone, the cost of government-sponsored farm subsidies increased from $2.7 billion in 1980 to $25.8 billion in 1986.[56]

These cynical, vote-buying policies have a threefold effect. (1) They ensure that the huge farming lobby will continue to support the existing governments in the United States and Europe during elections. (2) They lead to the production of huge quantities of artificially cheap subsidized food. (3) Third World countries rely more and more on this cheap imported food since most of their land is now used to grow cash crops to pay back their foreign debt. Bananas, cotton, rubber, cocoa, sugar, coffee, and flowers are exported at low prices because of a world glut in the wake of shortsighted IMF policies that forced many debtor nations to grow food and crops for export. As a consequence, poor peasant farmers are starving or forced to buy imported food because little land is left for them to grow food on.

To summarize, in 1980 about $35 billion was transferred from the poor Southern Hemisphere to the rich Northern Hemisphere because of adverse trade terms, repatriation of profits, "brain drain" (migration of well-educated people), and interest payments. In fact, the real aid passes from the poor southern countries to the rich northern ones. The overall relation between population increase and food production appears grim—more people and less food.[57] The remedy lies only in the application of money, science, and political will by the rich and poor countries in concert. No person has a right to starve in this world; conversely, no person has a right to be rich on this planet while others starve. There is potentially enough food for everyone, and each human life—be it black African, Indian, Asian, or Caucasian—is as precious and sacred as the next.

GENERAL AGREEMENT ON TARIFFS AND TRADE (GATT)

The Third World is about to suffer yet another insult to its basic integrity organized by the transnational corporations of the United States and other rich nations. These corporations straddle the oceans and conduct their affairs and business in many, and sometimes most, countries of the world. The term *transnational* therefore seems to be more appropriate than *multinational*. Companies that began in, say, the United States grew and became multinationals as they extended their influence into several other countries, but they became transnationals when they moved into almost every country of the world. Transnationals tend to be headquartered in the United States, the United Kingdom, France, West Germany, Japan, or Switzerland. Two of the more striking examples are Coca-Cola and Pepsico. Transnationals in practice are impersonal, faceless organizations answerable to no one except their shareholders. Many of them have budgets larger than the GNPs of most countries. They seem to be bound by no definite moral principles, except the profit motive.

A classic transnational is described in Greg Crough and Ted Wheelwright's book *Australia: A Client State:*

> The International Telephone & Telegraph Corporation, ITT, is involved in hotels, insurance, finance, teleprinters, cosmetics, automotive parts, pipes, pumps, food, cellulose, timber, fire extinguishers, coal and land development, as well as major telecommunication companies. It employs 348,000 people, nearly two thirds of whom are employed outside North America, and as one of its annual reports said, it is constantly at work around the clock, in 67 nations on six continents, in activities extending from the Arctic to the Antarctic and quite literally from the bottom of the sea to the moon.[58]

Transnationals have enormous power to control and manipulate the population, consumers, and politicians. They have their own global intelligence and communications networks, they can afford to buy business favors (bribes) from politicians, and they run million-dollar public relations campaigns. They pay for the very best lawyers to counter any legal challenge.[59]

Hundreds of years ago, before corporations were invented, poor countries were invaded and colonized by richer, more powerful countries. Their people were subdued and controlled by armies, and their natural wealth in spices, gold, silk, opium, forests, and minerals were plundered and stolen by their wealthy occupiers. Millions were killed and slaves taken. Colonialism was often supplemented by gunboat diplomacy, and recently covert operations have been used to subdue difficult Third

World countries. Since 1945, the United States has intervened militarily in foreign countries on the average of once every eighteen months, primarily in order to control its markets and provide access to cheap labor or natural resources.[60]

Despite past repression, though, a global movement toward nation-state autonomy is gaining ground. After India achieved its independence from England, other colonial countries followed a similar path to freedom in Africa, Asia, and, occasionally, Latin America. Yet, for all their immense influence in many countries, transnational corporations are still not satisfied. They feel threatened by the rising nationalism and independence of many Third World countries, and they have initiated a brilliant round of negotiations aimed at bringing all other countries under their ultimate control. If these words sound too strong, let me elaborate.

The organization known as the General Agreement on Tariffs and Trade (GATT), established during the reconstruction period after the Second World War, has been chosen by the U.S. government and the transnationals as their vehicle to regain total sovereignty over most nations in the world. A round of talks under GATT auspices was initiated in Uruguay in 1986. The talks continue as I write this in February 1991.[61]

This latest round of GATT talks was considered by many transnationals to be necessary for the reorganization of the international economy and the solution of economic problems into the next century. All relevant meetings are conducted behind closed doors in absolute secrecy, and only the major industrialized nations of Europe, Japan, the United States, and representatives from the transnationals are invited to attend. The last are also present as "advisers" to their national delegations. This decision-making process is called the "green room consultations," after the wallpaper of the GATT director general's conference room in Geneva. GATT documents and decisions are restrictive and secretive, and the media are barred. All other countries, including Third World nations, are excluded from the talks, and nongovernment organizations (NGOs), which represent church groups and humanitarian interests, are also excluded.[62]

GATT basically aims to enhance the influence, power, and control of transnationals in the Third World. It would curb the right of governments to use their economies and tax dollars for the benefit of their people, obliging them, instead, to provide "space" for the transnationals. Any transnational commencing certain operations in a specific country would be free to expand its activities into any other area (say, from mining to agriculture). The property rights of these rich foreigners would take precedence over those of the nationals. The transnationals would be allowed to produce or import goods or services as they pleased and to decide which technology, if any, could be imported and used in a certain country. In other words, countries would become prostituted to the transnationals.[63]

These GATT regulations, if implemented, would reshape the existing international trading system to give transnationals total freedom to operate globally. If these U.S.-inspired efforts succeed, developing nations may be forced to decrease or eliminate the regulation of investments and the control of the operations of foreign companies on their soil. These corporations would effectively control mining, manufacturing, banking, transport, insurance, media, communications, advertising, wholesale and retail trade, auditing, and legal practice. The specific governments would also be forced to introduce laws protecting and enhancing patents owned by the transnationals and other industrial rights. If these international laws were enacted, people in the Third World would discover that they did not even own the rare plants and seeds that were discovered in their forests of origin and that could be put to commercial use. Farmers could not store seed for the next season or even breed their own cattle, because the transnationals would own the seed and the livestock, and the farmers would be obliged to buy these from them.[64]

At the present time, many universities in rich countries work hand in hand with corporations. Although the U.S. government sponsors or pays for most research and development, corporations benefit from much of the resulting knowledge. Since the Reagan years, almost all new scientific knowledge (even some medical) is classified because it has relevance to the military-industrial complex or because of corporate ownership or jealousy. Thus, much scientific research or knowledge is now classified and/or owned by corporations.

Fundamentally, the world's knowledge and technology base is the private property of the major transnationals. Research and development (R & D) conducted by approximately three million corporate-related scientists and engineers costs more than $150 billion. These figures amount to almost 70 percent of the world's scientists and approximately 85 percent of the world's expenditure on R & D. General Motors spends more on R & D than does the country of Belgium, and IBM more than doubles the amount spent by India with its one billion inhabitants. Transnationals hold the bulk of the world's patents, 90 percent of which originate in developed capitalist countries. In seventeen African countries, ten foreign companies owned 90 percent of the patents; in Australia, in 1980, 93 percent of the patents filed were foreign.[65]

The transnationals are the world's major sources of technology, and they control it legally through patents, investments, cross licensing, and services. At the moment, a huge chasm separates the knowledge-rich northern and the knowledge-poor southern countries, and it is about to widen, as transnationals jealously guard and protect their information, which they will under no circumstances share with Third World countries. This means that these handicapped nations will never develop or improve their standard of living and will remain poor and beholden to those in the Northern Hemisphere and the First World.[66]

At the same time, though, old redundant technologies are exported to the poor south, while the domestic markets of the rich north are closed to most manufactured or industrialized goods from the south. Toxic drugs, banned by the FDA, poisons, pesticides, weedicides, and other noxious chemicals are already freely dumped by transnationals into the Third World. In 1987, world pesticide sales totaled $16.8 billion, and almost half of India's agriculture is now treated with chemical pesticides (compared with less than 5 percent in the 1960s). Some two million people are poisoned each year by pesticides, mostly in the Third World, and the National Academy of Sciences says that pesticides may induce 20,000 cancer deaths per year.[67] The GATT agreement is threatening to allow foreign corporations even more access and to force Third World countries to accept the foreign agricultural products flooding their markets.[68] Compliance will be compelled by the threat of trade retaliation. The transnationals will be able to take and sell what they want while restricting the access of Third World goods to the rich First World.[69]

Great advantages will be reaped by agribusiness transnationals, which will take over vast tracts of Third World land to grow cash crops, while peasants will suffer and forests be destroyed. Of course, they will use ample amounts of toxic pesticides and artificial fertilizers to grow this food. Even now, three-quarters of the fruit and vegetables America consumes comes from the Third World.[70] Incidentally, these are the same agribusiness companies that took over the land when thousands of proud Iowan farmers became bankrupt during the agricultural recession induced by Reagan policies.

Under the GATT agreements, Third World countries will lose their ability to control timber exports, and it is predicted that the price of timber from tropical forests will be reduced from $6 billion to $2 billion by the year 2000. Fast-food hamburger chains are already urging the U.S. government to use "free trade" to abolish U.S. beef import quotas or tariffs, and this policy will obviously lead to increased destruction of the Amazon forest.[71]

Let me give an example of GATT's power, control, and philosophy. In August 1991, a tribunal of trade experts from Hungary, Uruguay, and Switzerland mandated that GATT prohibit its 108 members from imposing import restrictions founded on environmental issues. This policy would overthrow the U.S. ban on tuna from Venezuela, Mexico, and Vanuatu. Tuna from these and other countries, including Costa Rica, France, Italy, Japan, and Panama, were banned from the United States because their fishing fleets kill many dolphins during the tuna catch. GATT will have the ability to overturn many essential national environmental laws in the name of free trade. The U.S. Congress has just agreed to "fast track" legislation—that is, to endorse all of the GATT proposals as they appear—an extremely dangerous development![72]

Clayton Yeutter, chief GATT negotiator under Reagan, has said that GATT reforms are necessary to confront nations that use tough environmental regulations as "trade barriers," and Richard Darman, director of the White House Office of Management and Budget, said at Harvard on May 1, 1990, "The environmentalists are promoting an antigrowth movement—Americans did not fight and win the wars of the 20th century to make the world safe for green vegetables."[73]

So the term *free trade,* which the White House likes to bandy about, means "free trade" for the transnationals and "total trade restriction" for all countries opposed to the global exploitation by the corporations. Another favorite phrase describing GATT objectives is "a level playing field," which also means that the transnationals can exploit other countries with impunity.

GATT will affect national sovereignty, the environment, Third World people, farming, food and safety standards, and ownership of minerals and national resources. In other words, GATT will achieve by a new legal world order what colonialism, gunboat diplomacy, and covert operations could not—total transnational control of the resources of all countries. Because these corporations will control all domestic services as well, the only role left for Third World governments will be the maintenance of law and order and the control of labor and unions.[74]

Perhaps this is what George Bush meant in his 1991 state of the union message when he talked about "a new world order." I am also very worried about the future of the Soviet Union and the Eastern bloc countries at this time in history. Gorbachev opened up the Union of Soviet Socialist Republics to change, through the introduction of *glasnost* and *perestroika.* Although I am not surprised that chaos has occurred, because it took Britain hundreds of years of strife and war before it developed the stable form of government known as the Westminster system, or parliamentary government, I am alarmed that the East European countries are making themselves so readily accessible to "free trade." Already a McDonald's restaurant, the largest in the world, is operating near Red Square. In the year after opening day, it served fifteen million customers.[75] The transnationals move in where angels fear to tread—what a wonderful opportunity to enlarge their influence, their markets, their profits, and their power! They know that the Soviet Union has vast quantities of untapped natural resources, including oil, minerals, and forests. At the moment, unfortunately, the Soviet people see only the wonders of capitalism and not the sinister, greedy, and ecologically destructive behavior of the transnationals. Nor, I think, do they appreciate that the IMF, the World Bank, and the U.S. Congress and executive branch are controlled by these same mammoth corporations. With the Soviet Union in chaos, and with a new détente at hand, Bush's new world order implies, I fear, total transnational domination of all countries, including the USSR, Eastern Europe, and China.

China is another vast market in the making. On my visit in 1987, I came down to breakfast in the Great Wall Hotel, in Beijing, built by an American corporation. I already felt uncomfortable in this atmosphere of affluence, for the hotel towered over millions of dirt-floor hovels occupied by the people of Beijing. I encountered an American in the lobby and asked him uneasily how he felt about China. He settled back in his chair with one leg draped over the armrest and said, "Oh, there are great opportunities here." I nearly vomited. I had spent three weeks traveling around China, developing enormous respect for these industrious people and their ancient civilization. They had been able to extricate themselves from the clutches of British colonialism over the last thirty years and had managed to feed, educate, house, and clothe one billion people and had successfully dealt with their overpopulation problem. The revolution had some severe setbacks, as do all political systems, including the cultural revolution of the 1960s, with loss of civil liberties and, in 1989, the Beijing massacres. But what will happen to this revolution if the transnationals take over? They will almost surely bring the end of equality and the beginning of exploitation, divisions between rich and poor, and a reversion to colonialist times. What a tragedy if that should occur!

The world economy is in chaos with no one in charge, except the exploitative transnational corporations. In 1983, Pierre Trudeau, then the prime minister of Canada, told me about an economic summit he had recently attended in Europe. President Reagan greeted the heads of state and then said virtually nothing for the rest of the meeting. He used Margaret Thatcher as his front "man," and she so intimidated François Mitterrand, Helmut Kohl, and Bettino Craxi that they were rendered speechless. Trudeau said that none of the leaders displayed any real comprehension of global economic policies. They relied instead on discussion papers written by civil servants. The whole meeting was a farce. The participants dealt only with the economic realities of the industrialized nations; they made no mention of the Third World, that is, 80 percent of the earth's population.

We know who controls the U.S. government. In speaking to hundreds of thousands of Americans, I always ask, "Hands up those of you who believe you live in a true democracy." One out of two thousand people may raise a hand. I then ask, "Who runs your country?" and they answer, "The corporations do."

8

The Manufacture of Consent

In order to appreciate the influence and power the transnational corporations exercise in the Third World, we must understand how they obtained this power in the first instance. Many of the global transnationals had their origins in the United States as small businesses in the last century. By a process of competition, hard work, takeovers of smaller companies, and some ruthlessness, they became enormously wealthy and powerful.

How did they manage to gain the support and cooperation of the American people as they achieved their power? The truth is that they have not only been competitive and somewhat ruthless in the Third World but have also exhibited a history of exploitation and subtle coercion within their country of origin. Most Americans are not familiar with this history.

Although this "manufacture of consent," to use the phrase coined by the respected journalist Walter Lippmann, was investigated and soundly denounced by national church groups and congressional committees, the political and propaganda activities of U.S. corporations have continued unabated over the years and to this day remain largely unknown.

The story of corporate propaganda explains, I believe, the strange and powerful patriotism and nationalism of the American people, their ready acceptance of propaganda and media manipulation, and the reality that American workers are not represented by a broad-based, powerful union movement. It also explains why the minimum wage in 1991 was only $4.75 per hour and why there are insufficient occupational health and safety standards, no uniform free health care system, and no national system of free higher education.

By comparison, the people of my native country are quite blasé about nationalism or patriotism; we rarely fly flags. But the minimum wage is $10 per hour, and the union movement is strong. The workers have excellent occupational health and safety standards, and we have a free health care system and an almost free university system.

This chapter and the next will continue to examine the etiology of the global ecological crisis. This may be a difficult chapter for some people to read and digest because it challenges many basic beliefs of American society, but I hope you have the courage to look at the truth.

I have relied in the following discussion on the work done by the late Alex Carey, an Australian psychologist who lived in the United States during the 1970s and 1980s. He was fascinated by the origins of American culture and made an extensive study of the history of propaganda and its political effects.

From the beginning of this century, large-scale professional propaganda campaigns have been waged by American business in order to shape public attitudes to accept and endorse the capitalist system, and this propaganda has changed the direction of American society.[1]

The campaign began between 1880 and 1920 in Britain and the United States when the right to vote was extended from 15 percent of the adult population to 50 percent. This popular franchise immediately posed a threat to the rich minority, because as real democracy was instituted, people would naturally be voting for laws that supported their own health, education, and welfare. For the first time, their tax dollars would be used to support the majority of the population and not just the rich. In 1909, two leading scholars—Abbott Lawrence Lowell, president of Harvard, and Graham Wallas, a leading British student of democracy—warned that the consequences of those new laws might be dangerous. They said, "Popular election may work fairly well as long as those questions are not raised which cause the holders of wealth and power to make full use of their resources. If they do so, there is much skill to be bought, and the art of using skill for the production of emotion and opinion has so advanced, that the whole condition of political contests would be changed for the future."[2] In other words, if the power of the rich is challenged, they will use their money to intimidate and coerce people, to buy votes, and to produce a mandate for themselves.

By 1913, a congressional committee was established to investigate the activities of an organization called the National Association of Manufacturers (NAM), which represented many U.S. businesses and had already begun disseminating vast quantities of literature with the apparent intention of "controlling" public opinion in the fledgling democracy.

But public opinion moved away from and did not support corporate philosophy during the First World War, which ended in 1918, because the American people were encouraged to work together for the good of the country. Women took equal jobs side by side with men, and a mood of unselfishness and generosity prevailed in the country.

PROPAGANDA TECHNIQUES

The techniques of propaganda were developed during the First World War when the American people were reluctant to become involved in the war because at that time they had no specific animosity toward the German people, although Germany certainly antagonized many Americans when it sank the British passenger liner *Lusitania,* with the loss of 128 American lives. President Wilson and others initiated a large and very effective campaign, under the supervision of the Committee on

Public Information, headed by George Creel—a Denver newsman—to convince the nation that Germany, whose people were called Huns, was the seat of all evil.[3]

Propaganda is "the organised spreading of ideas, information or rumour designed to promote or damage an institution, movement or person,"[4] and the First World War marked the first time in history that propaganda had been successfully conducted on a large scale. Within six months, the American people were devoted to hating the Germans and to defeating them in the war effort. (Does it sound familiar? Replace Germany with Iraq.) Public opinion at that time had been so aroused that grotesque campaigns of witch-hunting and Americanism abounded.[5]

One of the creators of propaganda during the war was Edward Bernays, a nephew of Sigmund Freud, whom he closely resembled. (I met him in 1981 on a cold rainy Boston day, and he offered to help in my anti–nuclear war campaign. In the end, he was unable to contribute, but I learned some rather remarkable facts. As we sat drinking tea, looking over a narrow Cambridge street, he told me proudly that he was the person who taught women to smoke, by dressing them in beautiful clothes, placing a cigarette in their hand, and adorning *Vogue* magazine with their photographs. I felt ill.) Bernays headed the transfer of wartime propaganda skills to the business arena. When the war ended, Bernays wrote, business "realized that the great public could now be harnessed to their cause as it had been harnessed during the war to the national cause, and the same methods would do the job."[6]

Incidentally, in 1919, Samuel Insull, who headed a vast utilities empire, took over the entire hierarchy of the "information committees" set up by President Wilson during the war, and he used them to great advantage to protect private utilities against the threat of public regulation or ownership. Here lie the seeds of the public's remarkable support over the years for private utilities and for nuclear power, advocated by these utilities.

After the war, propaganda was recognized as a tool corporate America could use to further its own agenda. In 1919, some 350,000 U.S. steelworkers went on strike and demanded shorter working hours and higher pay. (They were working eighty-four hours per week.) Because they had felt proud of their work during the war, they naturally expected that they would be well treated. But they were not. Instead, the U.S. Steel Corporation bought full-page advertisements in newspapers to encourage the strikers to return to work and to accuse the strike leaders of being Bolsheviks and Reds (at a time when the Russian revolution was in its infancy) and of being Huns. They also told the American people that the price of steel would soar if the workers prevailed. How could the U.S. corporations call proud American workers Communists and Germans when they had rallied so staunchly behind the war effort?

Unfortunately, a gullible public was largely persuaded by this propaganda offensive. The strike was defeated by the artificial engineering of public opinion, which had been successfully turned against the workers. By the time the strike ended, twenty workers had been killed, the hours and wages remained the same, and the price of steel went up. So the steel industry won.

Organized unions, I believe, are the best and only vehicles for the representation of the true interests of the working people of America—health care, occupational and safety standards, wages, working hours, and so on. They constitute the sole force that can take on, and to some extent control, corporate power. But since 1919, U.S. corporations have systematically worn down, demoralized, and destroyed organized labor by using the techniques of propaganda, Red-baiting, and intimidation.

PUBLIC OPINION

The successful propaganda campaign that ended the steel strike was then extended to American public opinion at large. Corporate America started the rumor that American workers and their leaders wanted to overthrow the federal government. It introduced this unsubstantiated notion through a public relations campaign in the media that led to an intense period of virulent anticommunism, in the years 1919–21. The witch-hunts and blatant Americanism also continued after the war, and people were actually jailed for practicing their right of free speech.[7] As a result, many American citizens felt persecuted and alienated within their own country. This propaganda campaign proved to be very effective, and the American public was persuaded to support the rights of rich citizens and corporate power, while support for civil liberties, social reform, and the labor movement declined.

In the 1920s, Edward Bernays called this use of propaganda "the engineering of consent," and Harold Lasswell, for fifty years the leading American scholar of propaganda, said in 1939 that propaganda had become the principal method of social control. Lasswell remarked, "If the mass will be free of the chains of iron, it must accept the chains of silver. If it will not love, honour and obey, it must not expect to escape seduction."[8] In other words, if ordinary people gain power in a democracy through the vote, then the rich will find another way to maintain control.

Everything went well for corporate America during the 1920s, but the country suffered during the Great Depression of the 1930s. Tens of millions of people lost their jobs, the banks collapsed, and people starved. Among the poor and indigent arose a great wave of hostility and animosity directed toward big business and corporate power, and people demanded a more equitable distribution of wealth.

In 1932, Franklin Delano Roosevelt was elected president, and his administration launched the New Deal, which cared for and gave succour to millions of unemployed, depressed people. Mass employment schemes were initiated to rebuild cities, bridges, and roads, as were other public works. The common people of America loved and trusted FDR, many being mesmerized by his "fireside chats," broadcast on the radio. It became morally and politically acceptable to advocate government ownership, government programs, and socialism as such.

But business people never liked or accepted President Roosevelt, because they had temporarily lost the public's loyalty, so they set out once again to recapture the minds of the people.

How did they do this? Well, they spent millions of tax-deductible dollars on public "education" programs and on polls. They also taught "human relations" to their own workers in order to control their thinking.

In 1935, the by then renowned National Association of Manufacturers (NAM) organized another massive propaganda campaign. The president of the NAM told business leaders in 1935, "This is not a hit or miss program. [It is] skillfully integrated . . . to . . . blanket every media. . . . It pounds its message home."[9]

In 1939, the La Follette committee of the U.S. Senate reported that the NAM had blanketed the country with propaganda that relied on secrecy and deception. The NAM employed radio speeches, news cartoons, editorials, advertising, motion pictures, and many other propaganda techniques that did not disclose its sponsorship. One business-sponsored agency distributed a steady supply of canned, ready-to-print editorials to twelve thousand local newspapers, and some 2.5 million column inches of this material were published.[10]

POLLING

By the late 1930s, public opinion polling had been invented, and it turned out to be highly useful to business. It was employed, according to Alex Carey, as an "opinion sensitive radar beam," which continually assessed ideological drift in the population. The polling data were used by the industrial propaganda institutions to provide continual flow of data and feedback, so that they could define and redefine their probusiness messages to make them more effective. It was also used to evaluate public response to product marketing.

NATIONALISM

In 1945, the corporations invented a new method to sell their capitalistic philosophy, which they called "techniques for community ideas." They discovered that the American people were not very excited by the rather sterile concepts of capitalism or free enterprise but that they did exhibit a rather positive emotional response to the notion of "Americanism." From this new information, the corporations devised a formula that tied many fundamental values together:

free enterprise = freedom = democracy = family =
Christianity = nationalism = God.

The equal and opposite formula they devised went something like this:

egalitarianism = equality = government interference =
socialism = unions = communism = Satan

These two formulas became the backbone of corporate philosophy and profit-oriented activities and propaganda, and they have been used ever since with undiminished success.

Just watch TV carefully for one night, and you will understand what I mean. A big, tough, hairy guy hefts a can of beer, the ad implying that all red-blooded Americans drink this brand of beer. The scene fades with patriotic music in the background. There may even be a suggestion of an American flag waving in the background. Advertisements also imply that freedom and democracy are God given, that the family is the fundamental unit of American society, and that all is well with the world, if you buy the products in the ads.

STRIKEBREAKING AND PUBLIC RELATIONS

In 1937, a steel strike erupted at Johnstown, Pennsylvania, when Bethlehem Steel refused to acknowledge the steel union. At that time, the corporations needed to gain control of a restive population, after the years of the Depression and the New Deal. So the local chamber of commerce joined with the NAM and Bethlehem Steel to orchestrate another propaganda campaign using the steel strike as the fulcrum. The National Citizens Committee was organized and launched by local businessmen; it engaged an advertising agency and a public relations council. The committee broadcast its antistrike messages of "Americanism" twice over a national network, and two full-page ads appeared in thirty newspapers in thirteen states. The campaign was once again successful.

At the end of the strike, James Rand, of the Remington Rand Corporation, proudly announced, "Two million businessmen had been looking for a formula like this, and business had hoped for, dreamed of and prayed for such an example as you have set."[11] This antistrike tactic was called the Mohawk Valley formula, and since that time this scientific strikebreaking technique has been used in every major strike in the United States.

The Senate-based La Follette committee criticized the propaganda tactics of the NAM in the 1930s, building up to the 1940s, in the following way: "The leaders of the association resorted to 'education' as they had in 1919–21. They asked not what the weaknesses and abuses of the economic structure had been, and how they could be corrected, but instead paid millions to tell the public that nothing was wrong and that grave dangers lurked in the proposed remedies."[12]

But the NAM continued to use its propaganda campaigns in the fight against labor unions. The corporations fought the most important strikes in 1945–46 in the press and over the radio, not in the picket lines. In effect, business owners bypassed the workers and went over their heads to appeal to the public, using false and unfair statements.

Fundamentally, these corporate campaigns were designed to achieve three objectives: (1) to minimize wage rises and to maximize profits, (2) to oppose decent working hours, a minimum wage, occupational health and safety standards, and employee health coverage, and (3) to prevent government regulations from interfering with their activities.

Over the years this corporate philosophy prevailed, so the workers of today are almost totally unprotected, with the minimum wage set at only $4.75 per hour. By comparison, unions in Australia have always been very strong, and as a result Australian workers have a decent minimum wage of approximately $10 per hour and are protected with good health care systems and safe working conditions. By and large, the Australian public and businessmen are tolerant of strikes because they understand that civilized negotiations will always be conducted and everyone will eventually benefit. However, U.S. corporate propaganda is now starting to infiltrate Australia and to affect its union structure. It is imperative that Australians learn from the past mistakes of American society, before our relatively compassionate society is degraded and changed forever.

The propaganda barrage by U.S. corporations continued through the years 1946–50. During this time, the NAM distributed 18,640,270 pamphlets that vehemently pushed anticommunist, antisocialist, antiunion, and anti–New Deal sentiments versus free enterprise and capitalization. Some 41 percent of these pamphlets were sent to employees, 53 percent to high school and college students, and 6 percent (over one million) to community leaders, such as ministers and women's club leaders.[13]

In 1946, the U.S. Chamber of Commerce distributed a million copies of a fifty-page article entitled "Communism in the United States." In 1947, a similar distribution occurred for a pamphlet entitled "Communists within the Government," which alleged that about four hundred Communists held important positions in the government.

The most effective propaganda weapon for both employees and college students was found to be a comic booklet. The *National Association of Manufacturers News* of February 1951 proudly proclaimed, "If all NAM produced pamphlets ordered for distribution to employees, students and community leaders in 1950 had been stacked one on top of the other, they would have reached nearly four miles into the sky—the height of sixteen Empire State Buildings, a record distribution of 7,839,039 copies."

The American Advertising Council, which was established fourteen days before the United States entered World War II to combat people's enthusiasm about the New Deal and their disaffection with the free-enterprise system, represents large corporations and advertising agencies. In 1947, it announced a twelve-month, $100 million campaign, one of numerous related campaigns to "sell" the American economic system to the American people. Daniel Bell, a professor of sociology at Harvard, said in 1954, "The output is staggering. The Advertising Council alone in 1950 inspired 7 million lines of newspaper advertising stressing free enterprise, 400,000 car cards, 2,500,000,000 radio impressions. . . . By all odds, it adds up to the most intensive 'sales' campaign in the history of industry." The campaign was used to "rewin the loyalty of the worker which now goes to the union and to halt creeping socialism, with its high tax structure and quasiregulation of industry."[14]

Fortune magazine in September 1950 carried an editorial saying, "The Free Enterprise Campaign is shaping up as one of the most intensive 'sales' jobs in the history of the industry—in fact it is fast becoming an industry in itself."

This deluge of brainwashing paved the way to the shameful era of McCarthyism in 1950–54, when Senator Joseph McCarthy intimidated, hounded, discredited, shamed, and destroyed the lives of hundreds of his fellow American citizens, until finally one man, Joseph Welch, was brave enough to confront him in a Senate hearing and speak the truth. Soon thereafter McCarthy died. From 1950 to 1965, corporate power was again safe from the threat of a freethinking skeptical democracy.

In 1955, *Fortune* magazine estimated that there were five

thousand U.S. companies supporting public relations departments, at an annual cost of about $400 million.

Do you see how the cold war was a logical spin-off from these successful initiatives to control domestic thinking? The somewhat contrived threats of Russia and communism, not innocuous by any means, of course, were used primarily to intimidate and silence the freethinking working people of America, but the deadly nuclear arms race was a result of this domestic manipulation.

DESTRUCTION OF PRICE CONTROL

The same tactics combining public relations with corporate political ideology were also used by big business to obtain uncontrolled price rises. The campaign I am about to describe was a prototype of techniques that continue to be used to "manage" democracy in the interests of American business.

After World War II, President Harry S. Truman was worried about rising prices because goods were in short supply. He decided to keep prices low by supporting and extending the life of the federal Office of Price Administration (OPA), which had been established during the war.

But business wanted prices to rise. Represented once again by the NAM, it therefore launched a massive campaign against the OPA by printing millions of leaflets that were stuffed into the shopping bags of housewives. It also published full-page ads that stated cleverly but falsely that price controls themselves were the cause of the goods shortage.

In 1946, the Opinion Research Council, monitoring the results, found that at the beginning of the propaganda campaign 81 percent of the American people favored OPA but that at the end only 26 percent supported a continuation of the OPA. Americans had been manipulated yet again to act against their own best interests.

A discouraged Truman said, "Right after the end of the war, big business in this country set out to destroy the laws that were protecting the consumer against exploitation. This drive was spearheaded by the NAM."[15] In the effort to kill the OPA, the NAM spent $3 million, which in those days was a lot of money. As a direct result of this operation, consumer prices rose 15 percent and food prices 28 percent between June and December 1946. The people had again been exploited.

To add insult to injury, these price rises canceled the wage rises that labor had obtained from some of its more successful 1946 strikes; at the same time, real wages dropped from $32.50 per week to $30.00 per week, while yearly corporate profits reached their highest point in history, $12 trillion, 20 percent higher than those of the best war year.

It is interesting that since 1918 the Soviet government brainwashed its people by consistently lying to them, but its techniques were so clumsy that the people knew they were being brainwashed. By contrast, in the United States, corporations became expert manipulators, so most people have swallowed the corporate doctrine whole.

HUMAN RELATIONS—INDUSTRIAL PSYCHOLOGY

The corporations developed another nifty trick to convert their workers from "unionism" to "corporatism." It occurred to them that since most U.S. workers were captive audiences in their factories, if they appealed to them the right way, they could win their hearts and minds. Psychologists might be interested to know that the human relations movement was pioneered by corporate America for an ulterior motive. *Human relations,* a euphemistic phrase, was also called "employee participation," "employee communication," and "democratic decision making."

During 1945–46, business firms invested huge amounts of money in the study of this psychological discipline, and a plethora of books and literature appeared on the subject. Psychologists and social scientists were recruited to develop new and more effective methods to include workers in a science called "interpersonal communication," which was really used to induce workers to support their corporate bosses. By 1950, management had become obsessed with employee "communication." *Fortune* magazine noted, "There is hardly a business speech in which the word is not used."

These techniques of worker manipulation proved to be a successful tool for bypassing union power in the factories, and worker loyalty swung to management. In 1959, Peter Drucker, who represented American Management Consultants, said of human relations policies, "Most of us in management have instituted them as a means of busting the unions. That has been the main theme of these programs. They are based on the belief that if you have good employee relations, the union will wither on the vine."[16]

The literature on human relations continued to grow. During the decade of the fifties, there were four times as many studies of small human relations groups published in social science journals as in all previous publication history. Surprisingly, few sociological studies document the impact of this movement.

THE REAGAN ERA

For fifteen years after Joseph McCarthy died, the American public was once again placid and under control. Most people worked hard for their friendly corporation—it was like one big family where loyalty reigned supreme. But then came Vietnam and the civil rights movement, flower power, Woodstock, and Watergate, and the nation once again lost respect for corporate control. We must remember that 20 percent of the 250 million people in 1985 owned 44 percent of the money in the United States and that vast disparities between the very rich, on the one hand, and the middle class and the poor, on the other, must be maintained at all costs.[17] How else could the Mellons, the Rockefellers, and the others have become so hugely rich without this brilliant control of a so-called democracy?

In 1975, the Advertising Council therefore launched another

"economic education" campaign on the U.S. public. Two years later, *Fortune* described the council's continuing campaign as "a study in gigantism." By 1978, according to a congressional inquiry, U.S. business was spending $1 billion a year of tax deductible money for "education," to convince people that big government was bad for them. (The truth is that government regulation is bad for corporations. It is amazing how corporate advertising can turn truth on its head.) The campaign was once again successful, and public support for the proposition that government is bad rose from 42 percent in 1975 to 60 percent in 1980. On the coattails of this vastly expensive propaganda exercise was elected the doyen and figurehead of right-wing corporate America—Ronald Reagan. What an incredibly successful campaign! I remember it well. In 1975, the concept of big government was not a topic of discussion; by 1979, TV reporters would ask me, "But isn't big government bad?" I had no idea from where this concept had come. Now I know!

TREETOPS PROPAGANDA AND THINK TANKS

Brainwashing entered a more sophisticated phase in the 1970s. Until then, the propaganda offensives had been "grass roots," but now the corporations decided to establish a series of "think tanks" staffed by brilliant, erudite people who produced editorials, TV news pieces, and legislative material that was easy to understand, well conceived and written, and very acceptable to both the media and Congress. The material has always been provided in a timely fashion to guide legislation on a particular issue. This sophisticated, high-level manipulation is called "treetops" propaganda. Instead of being directed toward the man in the street, it is focused on influential decision makers in Congress and in the media—newspaper editors, columnists, and television. Its immediate purpose is to set the terms of debate and to determine the questions and agenda that dominate public discussion.

Here are a few instances of the terms of debate:

- In the wealthiest country on earth, should unemployment be maintained at 6 percent or at 10 percent? Not, should unemployment at any level be unacceptable? (Unemployment is good for business because it weakens unions' negotiating power by providing a pool of unemployed workers.)
- Should private doctors have more control over the medical system so that doctors make more money and only the rich get good treatment?
  Not, Does every person have a right to free state-of-the-art treatment?
- Is it economically desirable to eliminate CFC gas, should CFCs be reduced to 50 percent production by 1995, or would business lose too much money? Not, Should CFCs be eliminated completely?
- Would auto companies suffer too much if they made fuel-efficient cars? Not, Are fuel-efficient cars a necessity for saving the planet?

These think tanks are involved in "policy research" or "agenda setting" for the corporate benefit. Their goal is not to save the earth or to care for the American people but to enable the rich to get richer and maintain their power. I find it extraordinary that the rich expend so much effort and energy to gain ever more money and power, for these assets do not by themselves lead to happiness.

Although some private think tanks, such as the Conference Board and the Hoover Institute at Stanford University, have existed for several decades, some new, aggressive right-wing tanks producing an incessant flow of market-oriented studies were established in the 1970s. Among them are the Heritage Foundation, the American Economic Institute for Public Policy Research, the American Enterprise Institute (AEI), the Georgetown Center for Strategic and International Studies, and the Business Roundtable. Funders include such reputable corporations as Reader's Digest, Hertz, Coors, Holiday Inns, Ocean Spray Cranberries, Bechtel, Gulf Oil, Vicks (makers of Vapo-Rub), Amway, Hunt Oil, and the Chicago Tribune Company.[18] (Of course, there are also a number of think tanks that might be described as left-wing, among them the Brookings Institution, the Institute for Policy Studies, and the World Policy Institute, but they exert little influence on the public agenda.)

These think tanks virtually created the new conservative movement of the 1970s and set Reagan's agenda. The Heritage Foundation drew up a comprehensive list of agenda items for his first and second terms of office. The first document was called "Mandate for Leadership—Policy Management in a Conservative Administration." During his eight years in office, Reagan withdrew financial support for the United Nations (until recently the United States still owed several hundred million dollars to the UN in back debts from the Reagan years); undermined the trade unions; mined the National Parks; decreased funds for education, medicine, job training, community development, the poor, the elderly, and the indigent; stacked the Supreme Court with conservatives; built more nuclear weapons to develop "superiority" over the Soviet Union; attempted to create the capacity to fight and "win" a nuclear war; gave tax breaks to the rich and increased the sales tax; and undermined the Civil Rights Commission. But the actual agenda of these corporate-funded think tanks is to *(a)* decrease government regulation of big business, *(b)* decrease taxes for corporations and for the rich, *(c)* destroy the unions, and *(d)* increase profits.

In 1977, the AEI produced fifty-four studies on right-wing agenda items, twenty-two forums and conferences, fifteen analyses of important legislative proposals, seven journals and newsletters, and ready-made editorials sent to 105 newspapers. Public-affairs programs were carried by three hundred TV stations, and display units were produced for three hundred college libraries.[19]

The Business Roundtable, founded in 1972, comprises 197 chief executive officers from America's largest corporations. In financial terms, the total revenues of these companies represented in 1981 was equal to about half the GNP of the United States, or more than that of any other country in the world.[20] In 1972, Justice Lewis Powell, a Nixon appointee to the Supreme

Court, urged business "to buy the top academic reputations in the country to add credibility to corporate studies and give business a stronger voice on the campuses." This happened. In the 1970s, business established chairs of "free enterprise," filled with handpicked candidates, in forty colleges. This is a prostitution of classical education![21]

The Roundtable maintains a statesmanlike image, but according to Ralph Nader "the dominant purpose leading to the foundation was a desire to combat and reduce union power," and "it proclaims moderation while sabotaging moderate reform."[22] Although the Roundtable specializes in treetops propaganda, it also works closely with the NAM and the U.S. Chamber of Commerce in their grass-roots activities. Together, they in 1978 defeated labor law reform, which was established to help reinforce America's declining labor unions, and they worked to oppose important consumer protection bills.

To this end, the Roundtable hired a public relations firm that distributed canned editorials to 1,000 daily papers and 2,800 weeklies, along with cartoons that attacked consumer protection bills. It also utilized a fraudulent poll claiming that 81 percent of all U.S. citizens opposed consumer protection when independent polls showed that one out of every two people favored it. This poll was published in a full-page advertisement in the *New York Times*. According to *Fortune* magazine, the defeat of this bill was a signal of victory; in retrospect, it marked a watershed in the history of consumerism (its fate mirrors the defeat of Harry Truman's Office of Price Administration in the 1940s).

So the business of America is business—as demonstrated in the successful campaigns of 1919–21, 1946–50, and 1976–80. To quote Alex Carey, "Complete business hegemony over American society was established. On each occasion similar, if not more sophisticated propaganda and public relations techniques were used." But what about the ethics of this serious situation?

The Catholic church has always said that if you capture people's minds and hearts in the cradle, you have them to the grave. It is the same with American business. People are born into this corporate atmosphere, and they die embracing it. This loss of individual freedom occurs in the greatest country on earth, where "freedom" reigns supreme. I once asked a young woman why she thought America was the greatest country on earth, and she said, "Because we're free." When I asked her, "Who runs the Congress?" she replied, "The corporations." So I pressed, "What do you really mean?" and she answered, "We have the freedom to get rich." I suppose this is partly true, because the nation is, or was, so extraordinarily wealthy. Money does trickle down from the very rich, so most people can afford one, two, or three cars, refrigerators, houses, and so forth. Even the poor usually have a refrigerator and a car, but not the schizophrenics and alcoholics on the streets of New York. Yet, despite the overall appearance of affluence, most people cannot afford to be sick. Even those who are covered by private health insurance are not adequately covered for a major illness; when they die, they leave their families destitute. This is discriminatory.

U.S. RECESSION AND DEBT

What has happened to the United States as a result of this social engineering? The dynamics are complex, but in a certain sense the Reagan years saw the culmination of unbridled corporate activities.

The truth is that America is not really a supremely rich country any longer. The foreign debt is, at $3.5 trillion,[23] larger than it has ever been, and if Japan were not economically prepared to support the U.S. foreign debt, American society would collapse, along with much of the Western world. The main reason that the U.S. economy is so weak is that the Reagan administration spent more money than all past U.S. administrations combined. And most of it went to build weapons of mass destruction. In fact, Reaganomics led the United States into a deep recession, converting it from being the biggest creditor nation to being the biggest debtor nation in history.

The recession has been aggravated by the Savings and Loans

(S & L) scandal, which was instigated when the Reagan administration deregulated the S & L industry. Crooked dealers moved in and made a fortune by investing government-guaranteed money in junk bonds and dicey real estate deals. Many of the modern high-rise buildings in the Sun Belt—in Dallas, Houston, and other cities—were built with this money.

The S & L fiasco may cost the U.S. taxpayers almost one trillion dollars over the coming years.[24] One of its prime instigators was Charles H. Keating, who ran Lincoln Savings and Loan. His professed philosophy was "Always remember the weak, meek and ignorant are always good targets." Keating allegedly paid $1.4 million to five U.S. senators, including John Glenn and Alan Cranston, in bribes on behalf of Lincoln Savings and Loan. These men, dubbed the Keating five, were subsequently acquitted in a court of law. Keating has been charged with criminal fraud for the $2.5 billion failure of Lincoln Savings and Loan, and he also faces a $1.1 billion civil fraud suit filed by the U.S. government—the biggest suit in history.

Michael Milken has been jailed for similar activities. The company Drexel Burnham Lambert, which financed junk bonds, has been charged with a $6.8 billion lawsuit. Large numbers of ordinary people lost an enormous amount of money collectively—in many cases, their life's savings.[25]

You see, the Reagan era introduced a rampant "me now" generation of corporate cowboys who made millions or billions of dollars by defrauding other people. The U.S. banks are currently in serious financial straits, partly because of the Third World debt and partly because of unwise and unregulated lending to these corporate cowboys, who often did not provide the banks with collateral. When property prices collapsed during the recession, banks discovered record bad loans on their books. Because the U.S. government guarantees the deposits of failed banks, as well as of the S & L industry, it now faces a crisis in the banking industry that threatens to be as massive as the S & L fiasco. For example, one of the largest banks in America, the Bank of New England, collapsed early in 1991, costing the U.S. government $2.5 billion. Another similar collapse would obliterate the government fund that guarantees the banks; more are predicted in the near future.[26]

Meanwhile, to add fuel to the fire, the U.S. insurance industry is in deep trouble. Insolvencies are on the rise. The assets of the largest insurance companies are equivalent to the assets of the largest banks. In 1990, forty-four insurance companies failed, compared with thirty-one in 1989. The failures are attributed to unwise insurance company investments in real estate, junk bonds, and other areas. (The federal government does not regulate this industry, which at present controls $32 billion in real estate, $267 billion in real estate mortgages, and $65 billion in high-risk junk bonds.)[27]

The eight years of Reaganomics encouraged a kind of frenzy among American money-makers. People played the stock market, buying and selling junk bonds. They invested in futures markets, they indulged in insider trading, which was illegal, they pursued acquisitions and mergers on a grand scale, they set up cross-directorships, and they gambled with shareholders' money. They also paid huge salaries to themselves and corporate executives. Many of these activities were illegal, and only now are the federal government and law authorities catching up with these gamblers and cowboys. These people have seriously jeopardized the banks that lent them money, they have threatened the U.S. insurance industry, and they have destroyed the S & L industry.

The hectic days of leveraged buyouts financed by junk bonds have ended. Over the year 1990, company mergers and takeovers fell by one-third throughout the world and by one half in the United States, according to the U.S. firm Securities Data.[28]

The federal deficit is spiraling above $300 billion a year (equal to the annual military budget), and the Persian Gulf war cost more than $1 billion a day. Alan Greenspan, chairman of the Federal Reserve, warned that the present economic downturn or recession could become severe if the gulf war continued beyond April. Fortunately, it did not, and the allies contributed $48 billion to help pay for the war.[29] If the price of oil rises in the future, the situation will be aggravated.[30] The recession has caused the steepest decline in orders for durable and consumer goods since 1958. The deficit for the fiscal year 1992 is predicted to be $362 billion, and $115 billion of it to be used to pay back S & L debts.[31]

Yet while the American economy is under siege, the Coca-Cola Company announced a 13.2 percent earnings increase in the first quarter of 1991, to a record $320.9 million.[32] Fancy making this sort of money from the sale of colored candy water in plastic bottles! Where are our values, and what is the world coming to? Most multi- or transnationals continue to do well despite the federal government's struggle to repair the economic shambles left as a legacy by corporations that took advantage of the lax federal regulations introduced by the Reagan administration. This situation is a perfect example of the end result of right-wing think tank policies. The Reagan presidency marked a bonanza for big business and a tragedy for America, and who pays? The middle class and the poor—80 percent of the U.S. population.

PATHOLOGY INDUCED BY THE CORPORATE ETHIC

Insidious corporate proselytizing has inculcated into American workers an ethic that says, "Ask not what I can do for my country; ask what I can do for my corporation." But in a certain subliminal way, the sense of country and the sense of corporation have been turned into one and the same. If people have been convinced that these corporate values are the norm, they do not expect that they, as individuals, have an inalienable right to live in a society of compassionate values where free health care, free higher education, and a fair tax system are universal. But people in countries such as Australia, New Zealand, Britain, France, Canada, and those of Scandinavia experience these benefits and indeed take them for granted.

It is also becoming quite clear that most people in the States realize that something is very wrong when people die in parking lots of hospitals because they have no health insurance, or when families become almost destitute trying to provide a college education for their children. It is also obvious that while the rich and corporations pay little if any tax, the middle class and the poor bear the financial burden of running the country.

In fact, during the Reagan era, which brought the culmination of years of corporate planning, the rich gained and the poor lost ground, because Reagan decreed that tax cuts for the wealthy would stimulate investment and expand the economy and that the middle class and poor would benefit when the profits eventually "trickled down" to them.

Of course, this did not happen. According to the Congressional Budget Office, the top 5 percent of all Americans received 45 percent more income before taxes in 1990 than they did in 1980, but during this time their tax rate decreased by 10 percent, from 29.5 percent to 26.7 percent. In contrast, the poorest 10 percent of the American people earned 9 percent less income in 1990 than in 1980, and their tax rate increased by 28 percent.[33]

A way to tax the poor indirectly, without actually saying so, is to tax everyday items like clothes, gasoline, cigarettes, and beer. I am always shocked when I buy a dress in the States for $50 and the salesperson at the counter says, "No, it's $60 with tax." In order to end this deceptive practice, the price indicated on the label should include the tax. We do not pay sales tax in Australia, yet the right-wing influences are trying to have it introduced. Of course, the poor and middle class are most affected by sales tax; the rich hardly notice it.

In 1990, Congress was deeply concerned about the enormous deficit and knew that it must increase taxes, despite George Bush's "read my lips" speech. But members of Congress are reluctant to tax the wealthy, and big business is in a good position because it funds their elections and lobbies them between campaigns by means of right-wing think tank strategies and staff.

I pay 49 percent of my income in Australian federal tax, but I do not object and am even quite proud, because I know this money will be used to care for and educate me and my fellow citizens. But I am aware that people in the United States dislike paying taxes because they know that this money will generally not be used for their own benefit and well-being.

For instance, under the present tax structure, the people and businesses in the town of Lewiston, Maine, paid $44,720,000 to the Pentagon in 1990 and $45,272,296 to their city budget. A Lewiston family of four pays only $30 for elementary education, $33 for higher education, and $922 for the military, most of which goes to corporations that build weapons. Put another way, out of their average total federal tax bill of $2,144.90, about 43 percent goes to the Pentagon, 26 percent to pay interest on the federal debt, and 31 percent for all other spending on health, education, welfare, roads, and so on.[34]

Some 37 million Americans have no medical insurance, more and more middle class families can no longer afford the premiums, and the federal government does not provide free medical care. Ironically, despite these glaring deficiencies, the United States spends more on health per capita than any other nation. According to a Harvard University survey, the U.S. system costs more than three times as much as the Australian or the British system, each of which provides excellent health care for everyone.[35] If these countries can do it, so can America, which has always been innovative and creative as a nation.

Let me give a classic example of the attitude of corporate business toward the health care of its workers. Industry leaders have become increasingly concerned about mandatory employer-funded health care schemes, saying that they are unfair to the private sector and contribute to higher prices—everything is passed on to the consumer while business always profits. Lee Iacocca, chairman of Chrysler, said that the cost of employee health insurance added $930 to the cost of every Chrysler car.[36] Of course, the government should be paying for universal health care and not wasting $300 billion on weapons and the military-industrial complex.

During the Reagan years, large numbers of psychiatric hospitals were closed in order to "save money," and the streets of New York, Boston, and other cities filled with chronic schizophrenics and other mentally ill people, some of whom starved and froze to death in the cold. This policy was probably one of the Heritage Foundation's recommendations to President Reagan.

Private charities—those "thousand points of light"—are meant to care for these people when government abrogates its responsibility. Soup kitchens, churches, and good people are expected to take up the slack. Corporate America has for years encouraged the government not to provide social services for people. Instead, large numbers of tax-deductible charitable foundations have developed. They enable rich corporations to evade taxes by channeling money into charity and "educational" pursuits. Some of these foundations do actually care for needy people, but many of these fall between the cracks. It is tragic that charity is expected to care for America's poor, sick, and indigent, while tax dollars are pumped into the Pentagon, especially now that the Soviet Union is no longer intact and the Communist party has been outlawed by the Soviet federation.

In 1990, Congress tried to increase taxes for Medicare, which brought howls of rage from the elderly. Medicare payments by the public have been increased by more than 300 percent over the last ten years, six times the rate of inflation. Remember that Medicare provides little money for prescription drugs or for protection against financial ruin from the cost of nursing-home care. Many old people are therefore forced to deplete their life savings in order to pay for medical care. This is particularly tough for the elderly poor, for the 40 percent of those over sixty-five who live on an annual budget of $12,000 or less. This number includes nearly half the older women in America and two-thirds of the elderly minorities. As residents of the richest country in the world, these people deserve a measure of comfort and dignity in their old age.[37]

Australia, we noted, has a dual medical system: free medical care for all and a voluntary private health insurance scheme that enables individuals to choose their own doctor and private hospital. Acutely sick people are actually much better off in a public hospital with resident medical staff and excellent twenty-four-hour laboratory service. Private hospitals have nice rooms and nice food, but patients have been known to bleed to death at night without around-the-clock medical cover. In other words, our tax dollars are used for our care. I recently had my gall bladder removed by the best surgeon in town. I spent ten days in hospital and enjoyed excellent nursing care and tasty meals, and I paid not one cent. It was all on the government.

The U.S. medical system, by contrast, is almost totally privatized. I make these comparisons not to put America down but to suggest ways to foster a more compassionate society. The United States has created a large and advanced medical system capable of treating every known illness, yet people die in their homes because many cannot afford to see a doctor and be admitted to a hospital—a standard hospital bed can cost $2,600 for twenty-four hours, when the minimum American wage is only $4.75 per hour. This health care system concerns itself primarily with treating the rich. America is the only industrialized country, apart from South Africa, that fails to offer comprehensive medical care to all its citizens.[38]

Many hospitals are run by private corporations and entrepreneurs who see medical care as a profit-making venture. Nursing homes have become a huge "industry"—fancy calling the medical care of old people an industry, or do we make money out of just anything? The owners open and close hospitals at whim, depending on whether they are making money or not, hiring or firing nursing staff, admitting or discharging patients. Medicine should not be practiced for profit; it is a service that should be offered with humility and compassion.

I remember in 1969 when the president of the American Medical Association said that medical care was a privilege and not a right. This attitude violates the Hippocratic oath, a code of compassionate medical ethics—the very philosophy of medical practice. My first job as a newly qualified doctor was to relieve a country physician for two weeks while he went on vacation. He paid me several hundred dollars for my time, and I felt embarrassed to receive money for the privilege of practicing medicine. Medical care is a responsibility for doctors and the community. All people, clean or dirty, young or old, rich or poor, have a right to the best medicine that society can offer.

EDUCATION

Finally, I want to address the important subject of education, because to a certain degree the fate of the earth rests upon the education that children are now receiving.

In Australia, all universities, until recently, were funded by the federal government, and education for all disciplines, including law, medicine, architecture, and science, was totally free. Two years ago, the government introduced a small tertiary tax of $1,000 to $2,000 per year, which I fear may be the beginning of privatization of our university system. This new scheme is strongly supported and encouraged by right-wing think tanks employing the U.S. treetops philosophy and by corporate Australia. But my daughter, who recently graduated from medical school, had almost her entire education paid for by the Australian government.

High schools and primary schools are funded and run by the state governments, and education is free and uniform throughout each state. Like health care, free education is deemed a right of all children and people.

When I lived in Boston, I was surprised to discover that school districts in the state are autonomous and funded by the local population. It follows that, if schools are located in poor communities, the educational standards or facilities will be correspondingly low. In America, private schools are very expensive and available mainly to the middle class and the wealthy, as are many good colleges and universities, except for some excellent state systems like that of California. It is hard to believe that it can cost $20,000 or more to send one child to college for one year. For a family of six, educating four children breaks the budget and often leaves the parents in penury for years. Such a state of affairs is not fair, equitable, or right.

Another problem worries me. I speak regularly at hundreds of colleges across the United States and each year give several commencement addresses. I find that most college presidents are primarily fund-raisers who have little time to devote to the curriculum or education per se. Fund-raising is not an easy task; I know because I used to raise $3 million a year for my two organizations—Physicians for Social Responsibility and Women's Action for Nuclear Disarmament—from philanthropies. Colleges receive money from wealthy individuals and from foundations and corporate bodies, and their curricula are often influenced by the wishes and philosophies of the funders. This is not free Socratic education. Often when I am in a lobby of a college, I see a plaque over the door acknowledging generous gifts from IBM, GE, and other corporations, and I worry.

Apart from some of those at exclusive Ivy League schools, I find most college students to be almost totally ignorant about world geography, world history, English literature, the true history of America, or the history of the nuclear age. Most do not know about Hiroshima and Nagasaki or about Hitler, and do not have a true understanding of the etiology of environmental threats to the earth.

A National Geographic survey conducted two and a half years ago found that 11 percent of all Americans believed that they belonged to the Warsaw Pact, 50 percent could not name any Warsaw Pact nation, 56 percent did not know the population of the United States, and 50 percent did not know that the Sandinistas and Contras had been fighting in Central America. They really had little idea of the location, let alone the cultures, traditions, and attitudes, of most other countries in the world. The National Geographic Society said there is a state of "crisis" concerning the geographic literacy of eighteen- to twenty-four-year-olds in the United States. As one visiting Russian scholar said several years ago, "Americans have a curiosity a mile wide and an inch deep."[39]

Many of the American students I encounter have received little or no education in the ethics of business and often do not know what job they will get, except that it will be in the corporate world. I ask women who are graduating, "What will you do with your business degree?" and they say, "Oh, I want to go into marketing." When I ask further, "What will you market?" they say, "Oh, anything"—and that means really anything—

disposable diapers, CFC-propelled spray cans, cruise missiles. The engineering and mathematics graduates can find virtually no job except in the military-industrial complex, which has recently received a big boost from the Persian Gulf war.

These poor kids have not been given a comprehensive understanding of the world in which they live. If they are ignorant of the dynamics of the world, how can they protect their future? Furthermore, I find that they do not know how to think critically. When I give a lecture and tell them facts that are totally alien to anything they have ever heard before, they ask, "But how can we believe you?" and I say, "Don't believe me; do your own research and find the truth for yourselves." They seem perplexed and bewildered at the enormity of the task. They have not been taught how to use a library. I sense that most of them have been nicely conditioned and programmed to be conveniently pumped into the system of corporate America. Remember, it is the corporations in the Western world that are mainly responsible for global pollution, species extinction, and the threat of nuclear war. I am not excluding the dreadful mess the Communist regimes have made of the environment in Eastern Europe, China, and the Soviet Union.

Many college exams are multiple-choice and require little use of the English language. College teachers often despair as they tell me that students are poor spellers and have woeful math and reading skills. The teachers blame the primary and secondary educational system.

After the wild days of the 1960s when student sit-ins demanded a looser, more flexible system of education, the faculties capitulated. So now degree requirements consist of a semester of this subject and a semester of that. This educational scheme has produced several generations of Americans who are poorly educated in the affairs of the world.

I think it would be better if students were forced—and I mean forced—to study a particular subject in depth, so that they learn to use the five layers of neurons in their cerebral cortex. They really need to understand the concepts of critical thinking and research, and honest questioning—indeed, to develop a degree of skepticism toward what they watch on TV and read in the papers. Unfortunately, though, most articles in newspapers now resemble TV news, offering small "read bites" instead of "sound bites." I also find that most students do not read newspapers.

The standard of education should be uniformly high for all students, rich and poor, throughout the land. The federal and state governments must pay well to educate the next generation, and teachers must be free and independent of corporate and big financial interests. Some from this uneducated student body will emerge as the teachers of the next generation, thereby compounding the problem.

The ideal educational system parallels in many ways the ideal health care system. Only education is more vital for a nation. Our children are inheriting a dangerous world, dying from pollution and overpopulation, wired up by transnational corporations such as ATT, ITT, and GTE like a ticking time bomb ready to blow up at any minute from a nuclear war; yet they really have little knowledge of the situation, except for a vague understanding they obtain from watching TV news.

They must be well informed in order to ensure their own survival. Teachers, I believe, are the most responsible and important members of society because their professional efforts will affect the fate of the earth. I therefore think that they should be paid more than doctors, who, after all, only sometimes cure a sore throat or an acute appendicitis.

9

American Media and the Fate of the Earth

Let us take a searching look at the media, which are, we may safely say, determining the fate of the earth. If newspapers and TV decided to publicize the ozone destruction, global heating, deforestation, species extinction, overpopulation, and the threat of nuclear war as they publicize AIDS, they could educate, inspire, coerce, and persuade people to save the planet. Too often, though, we get only superficial or sensational news stories, such as those involving rape, murder, fires, and violence, which sell papers. Corporations who own the media are out to make money and to control the public through a perpetuation of ignorance, not to educate.

Yet at journalism schools, reporters are taught the ethics of journalism—fairness in reporting and thoroughness in investigative research. Often, good reporters and their articles are squashed by editors representing megamedia bosses like Rupert Murdoch.

In 1982, fifty corporations owned just over half the media business in America; by 1987, the count was twenty-six, and it is now down to twenty-three. A recent study conducted by the

International Labor Organization (ILO) predicted that a mere six conglomerates would control the world media by the year 2000. Big publishing and media companies are diversifying and taking over other companies. Their influence is spreading so rapidly that six megamedia giants ultimately control most companies that currently produce films, cameras, videos, tape recordings, compact discs, and books, as well as TV, radio, and newspapers. The study saw a pattern of vertical linkage developing between industrial producers, the advertising industry, and media giants. To put it simply, many of the important transnationals will be interrelated and will therefore work cooperatively. Such a situation is terribly dangerous. Economic control, political control, and mind control—that is just an extension of the GATT philosophy transferred from the Third World to the First.[1]

These conglomerates include Time Warner Inc. of the United States, Hachette of France, Rupert Murdoch's News Corporation, Canada's Thomson Corporation, and the Maxwell Communication Corporation in Britain.[2]

Apart from securing political and economic control, the media are used to make money. They do this by selling advertising, and if the corporate sponsors who buy the ads do not like the program content, they cancel the ads.

We must also examine the corporations that now own the U.S. media and their connections with other business. For instance, General Electric, which owns Raytheon, manufacturer of the Patriot missile, also owns the National Broadcasting Corporation. It is surely fair to ask, therefore, whether NBC could be impartial in its analysis and reporting of nuclear power stations, radiation accidents, demonstrations against nuclear weapons testing, the freeze, détente, or the Persian Gulf war. Impartiality appears impossible.

General Electric may serve as a prototype transnational corporation that has an enormous impact through the media. You might think that General Electric is true to its motto and brings "good things to life," like irons, stoves, washing machines, and refrigerators (all of which use electricity). But what is this corporation really doing behind its benign façade?

Its operations extend into fifty countries, in its search for markets, production facilities, and raw materials. Ronald Reagan was its devoted salesman for some ten years, as the host of the "General Electric Theater" television programs from 1954 to 1962. GE built an electric house for the Reagans in the 1960s, complete with such new inventions as a garbage disposal unit and a dishwasher.[3]

GE has been involved in nuclear weapons production since the end of the Second World War, as well as in the construction of nuclear power plants. In 1945, GE's president, Charles Wilson, opposed conversion of the military economy to civilian production and helped set in motion the machinery to ensure a permanent war economy.

Because of the early actions of Charles Wilson, GE had by 1991 become one of the largest nuclear weapons producers in the land, grossing $11 billion in nuclear warfare systems in the period 1984–86. It makes parts for the Trident and MX missiles and for the Stealth and B1 bombers. It is the developer and sole producer of the trigger for every nuclear weapon made in the United States; it manufactures Star Wars components, and it has a key role in the manufacture of all nuclear weapons (each one costs $40 million, and five new bombs have been manufactured every day), ranging from uranium mining, plutonium production, weapons testing, and nuclear waste storage.

Since 1945, GE has helped shape government policy to increase sales and profits for its nuclear weapons and related divisions. It has been very clever in these policy formulations, aided by a board of directors that has read like a Who's Who. David Jones is the retired chairman of the Joint Chiefs of Staff; William French Smith was Reagan's attorney general and is now his personal attorney. Other people associated with the GE board are Katharine Graham, owner of the *Washington Post;* Robert McNamara, former secretary of defense; Harold Brown, another former defense secretary; Cyrus Vance, Carter's secretary of state; and Alan Greenspan, chairman of the Federal Reserve. These and other board members sit on the boards of major U.S.

corporations like Quaker Oats, Goodyear Tire and Rubber, J. P. Morgan, and Citicorp.

GE executives also belong to key business groups and think tanks that exert enormous influence on government policy, including the Business Council, the Business Roundtable, and the Council on Foreign Relations. In addition, they are members of exclusive social clubs, where they fraternize with the elite and powerful—the Bohemian Club in California, the Economic Club and the Links Club in New York, and the Augusta National Golf Club. It is often within the confines of these clubs that some of the most important political decisions of the country are made. Not least, GE executives belong to very influential Pentagon committees. For instance, one executive who held various positions in GE, in 1987 headed a presidential space commission that strongly recommended that NASA develop a space station, and in that same year GE was awarded an $800 million contract for work on it. The chairman of GE, John Welch, was, until 1990, also chairman of the National Academy of Engineering. GE aggressively lobbies for its weapons systems. In fact, the company has more registered lobbyists than any other weapons contractor.

Like most corporations, GE has been involved in takeovers and purchases of other companies. For instance, it bought RCA for $6.4 billion in 1986, thereby also acquiring NBC.

Another network, ABC, is owned by Capital Cities, a huge company with interests in radio and publishing. In 1985, it bought ABC for $3.4 billion. But who owns Capital Cities? William Casey, deceased director of the CIA under Reagan, founded Capital Cities in 1954. When he was forced to put his stocks in a blind trust in 1983 because of his administrative appointment, he quietly kept control of his largest single stock, $7.5 million in Capital Cities.[4]

The history of the purchase of ABC is in itself very interesting. In November 1984, Casey, as CIA director, asked the Federal Communications Commission (FCC) to revoke all of ABC's radio and TV licenses because one of their news reports suggested that the CIA had attempted to assassinate a U.S. citizen. In February 1985, the CIA asked the FCC to apply the fairness doctrine to ABC; in March, Capital Cities bought ABC. Not a good beginning, for the newly acquired ABC, in impartiality and fairness in reporting. It was now owned, in effect, by the head of the CIA. Other board members of Capital Cities sit on the boards of, or are connected with, IBM, General Foods, Johnson & Johnson, Texaco, Avon, Conrail, and many others. See the interconnecting links between transnational corporations and the media?[5]

The powerful *Washington Post* was inherited and is now owned by Katharine Graham, who said to a group of senior CIA employees in November 1988, "We live in a dirty and dangerous world! There are some things the general public does not need to know, and shouldn't. I believe the democracy flourishes when the government can take legitimate steps to keep its secrets and when the press can decide whether to print what it knows."[6]

Without a free and uncensored press, there can be no legitimate democracy. If the people do not have the relevant facts, they cannot make informed decisions about their politicians, their country, or the world. Secrecy is promoted to maintain and not to threaten the power structure of the occasional woman and the white males who run the world. Many major decisions that affect our future are made behind closed doors, and then we are manipulated by the media to believe what we are told. The war in the Persian Gulf offers a good example of this strategy.

If society demands full financial disclosure from its politicians, it must demand the same from the corporations that own the media. Proposed legislation that would require disclosure of financial assets and ties is particularly pertinent when documentaries and stories are produced that could affect the profits of the corporate parent or the sponsors. These shows often do not go on the air or are not published. Furthermore, the actual ownership of a company must be made public. We tend to believe that we are dealing with a respected and well-known company, but because of corporate takeovers, the organization may really be owned by a bigger silent corporation, which probably cares little about the quality of the product but very much about profits.

It is no secret that Katharine Graham was close to the Reagans and that early in the presidency she had them to dinner in her Georgetown house. Thereafter she maintained close and frequent contact with Nancy Reagan by phone and lunch dates. The *Washington Post* also owns, with the *New York Times,* the *International Herald Tribune,* and it owns *Newsweek* and other large media affiliates. Its board includes the chief executive officers from Coca-Cola, H. J. Heinz, IBM, GE, and Prudential.[7]

The *New York Times* is 70 percent owned by the Ochs and Sulzberger families, but board members are also related to military contractors like General Dynamics, IBM, and Ford and to other blue-chip firms like J. P. Morgan, American Express, Manville, and others.[8]

And then there is the Public Broadcasting System, or PBS, which many people consider to be an impartial public network. But over the years it has become partly privatized by default. Many of the fine programs on PBS are sponsored by corporate polluters that are in trouble because they have added to the toxic woes of the world. These programs are used to redeem the image of the polluters. Because of these sponsors, PBS is facetiously referred to now as the Petroleum Broadcasting Service.[9]

Both PBS's redoubtable "MacNeil/Lehrer Newshour" and ABC's "Nightline," run by Ted Koppel, are seen as models of excellent, responsible investigative journalism. But Fairness and Accuracy in Reporting analyzed their guest lists and found some rather disturbing trends. Both programs interviewed a disproportionate number of white males. On the "MacNeil/Lehrer Newshour," 90 percent were white and 87 percent male. For "Nightline," the figures were 89 percent and 81 percent. The percentages were even higher when international politics were discussed. "Nightline" featured environmental issues in only 6 of 130 programs. Of the fifteen guests on these environmental shows, all were white and only two were women, one of them Margaret Thatcher. Nine of the fifteen were government officials and two corporate representatives. Only two were environmentalists. Ralph Nader said recently, "Look at all the stories on the destruction of the Amazon Forest. Do you ever see the names of any multinational corporations mentioned?" The forest destruction or other environmental disasters all seem to happen spontaneously, and no one is held responsible on these shows.[10]

When Robert MacNeil was asked why his program tilts toward the right, he replied, "There is no left in this country," and Jim Lehrer responded to suggestions that appropriate critics of government policies be given airtime by calling them moaners and whiners. On seven "MacNeil/Lehrer Newshour" segments on the *Exxon Valdez* spill, not one environmentalist appeared. Ted Koppel has said, "Policy critics aren't needed on Nightline since we invite the policy makers and ask them the 'tough questions.' "[11]

On these programs, women, half of the population, and people of color and union members are not represented. Once again a very small, powerful, unrepresentative section of the community dominates the airwaves, the debate, and the agenda.[12]

Hawks are put up to debate hawks. For instance, Caspar Weinberger is pitted against Senator Sam Nunn on the nuclear arms race. Why not Randall Forsberg (who invented the nuclear freeze concept and was one of the leaders of the peace movement of the 1970s and 1980s) and Weinberger, or Forsberg and Nunn? At least it would be a lively debate and, above all, informative.[13]

PBS has become more and more business oriented over the years, having fallen into the corporate orbit. It now runs the "Nightly Business Report," "Adam Smith's Money World," and "Wall Street Week." These programs are sponsored by Metropolitan Life and Prudential-Bache. Only 20 percent of the U.S. population owns stock, and only 5 percent buys or sells stock per year, so PBS is in many respects not serving the public. Rather, it is to a large degree controlled by and answerable to its corporate sponsors.[14]

Fairness and Accuracy in Reporting also conducted an important survey on union coverage. It sent a questionnaire to one hundred of the largest newspapers and compared their coverage of labor issues with their general business and economic coverage. At the same time, it reviewed the contents of the three nightly network news broadcasts and interviewed labor reporters, union representatives, and media professionals. According to the study, the 100 million American working people were continually ignored, marginalized, or misrepresented by the media. Workers were falsely portrayed as unproductive and lazy (often fat) and union leaders usually as corrupt. Unions were unfairly depicted as forcing corporations to pay exorbitant wages to nonproductive laborers. These higher wages were said to have led to decreased competition in international markets. According to the programs, unions always instigated conflict, and all unions were the same—corrupt.

The Machinists' Union also monitored TV coverage of unions and workers in the early 1980s and found that production workers were denigrated and that prostitutes were sixteen times more likely to appear on TV than mine workers, and fashion models ten times more likely than farm workers. Employers, by contrast, were portrayed as enlightened.[15] This scenario is the very agenda of the National Association of Manufacturers, the U.S. Chamber of Commerce, and the Business Roundtable.

Workers get a sympathetic press only in the context of environmental issues. If environmentalists try to save forests, the workers' jobs become paramount. Typical headlines are "Old Growth Forests versus Loggers," "Trees versus Jobs," and "Save the Planet versus Save Our Jobs."[16] The situation is similar in Australia. The forests must be destroyed to provide jobs for workers in the state of New South Wales, yet the government felt free to fire over a thousand railway workers when it cut back on rail services. No one wept for the sacked workers, but crocodile tears were shed for loggers.

It is interesting that the only workers in the world who receive good, sympathetic, almost laudatory coverage in the U.S. press are those in Poland's Solidarity movement and in Russia. So Communist workers are okay, but American workers are not. It all fits with the traditional hidden corporate agenda and the social engineering of America. Meanwhile, business and economic reporting received double the time given to workers' issues—health and safety, wages and unemployment—and strikes barely rated a mention or were reported on the back page.[17]

Interest rates, corporate managers, retail sales, trade deficits, and daily reports of the Dow Jones average are of interest to only a very small minority of American people. When I read the daily newspaper, it often seems so loaded with articles on finance, banks, profits, and corporate takeovers that I feel I'm reading a financial journal, when all I want is community, national, and international news. Perhaps it is all one and the same thing. After all, the corporations run the Congress, the White House, the Pentagon, the media, the banks, and the Third World. Soon they will economically control the Eastern bloc countries, the Soviet Union, and China—the whole world!

The Minneapolis-based Women against Military Madness (WAMM) conducted a survey on the tax-supported public radio station KSJN. It found that local peace and justice activists were largely ignored, while Reagan administration sources outnumbered other interviewees on "All Things Considered" by a margin of two to one and the noon news segment was dominated by people from commercial media and the military-industrial complex. WAMM said that this coverage was not representative of the tax-paying public.[18]

David S. Broder, the highly regarded journalist from the *Washington Post,* in a speech before the National Press Club in December 1988 criticized the cozy relationship between some journalists and the government. He cited the instance in 1980 when the journalists Patrick Buchanan, who once worked for Nixon, and George Will rehearsed Ronald Reagan for his campaign debate with George Bush. Broder called this "a breach of professional ethics so gross even Mr. Buchanan might be expected to grasp it." There is a growing tendency, Broder said, for journalists to dabble in politics and for government officials to enter the "revolving door" and emerge as prominent commentators and news executives. The First Amendment gives journalists a special immunity from government regulation, and this privilege must not be abused. The present situation, he pointed out, is very dangerous for the democracy.[19]

During the Reagan years, the White House press corps covered up for Reagan, knowing full well that he was not in control of the facts and that he could rarely speak extemporaneously, without the help of 3-by-5 cards or the Teleprompter. In 1983, I spent one and a quarter hours with him at the White House in private conversation about the arms race. His daughter, Patti, was the only other person present, and she said little. I judged his level of intelligence to be quite low. (The full description of that meeting can be found in my book *Missile Envy*.)[20]

Members of the press claimed that they initially tried to warn the U.S. public about the intellectual deficiencies of this so-called leader but that they received so many angry letters of support for the president that they stopped. However, the main task of the press is to tell the truth, not to be popular.

In the months before the 1984 presidential election, the press deliberately squashed several stories that could have damaged Reagan. ABC's special investigative unit had documented serious health violations at nursing homes owned by the director of the U.S. Information Agency, Charles Wick, a close friend of Reagan. It had also documented a White House and FBI coverup of Labor Secretary Raymond Donovan's alleged association with leaders of organized crime, and another story about the close Reagan ally Senator Paul Laxalt, who tried to stop the Justice Department probe into his campaign contributions. The stories were very good, but ABC editors spiked them.[21] Why?

On that note, Rupert Murdoch is notorious for electing and dismissing national leaders. He owns one-third of the newspapers in Britain, and his media definitely helped to elect Thatcher—they were her "cheer squad."[22] In Australia, where he owns 60 percent of the print media, he was a major player, together with the CIA, in the dismissal of one of our best prime ministers, Gough Whitlam, and he is a major supporter of our present prime minister, Bob Hawke. Murdoch makes a practice of interfering in the running of his newspapers, even after he has promised independence to his editors.

Peter Jenkins, a respected journalist who once worked with Murdoch in Britain, wrote that promises of editorial freedom "are of very little weight against a proprietorial or managerial ethos which is unfriendly to honest, fair and decent professional journalism." He added, "I had no cause for personal complaint against Murdoch, but I saw how good newspapers, and once independent spirits, wilted in his presence—or at 3,000 miles removed."[23]

Concentration of press ownership is extremely dangerous for a society. In the Soviet Union, the government owns the press, but in the United States, Britain, Canada, and Australia the press is owned, controlled and manipulated by a small coterie of extremely powerful, wealthy people. In a certain sense, these people control the Western world for themselves and for their buddies who own the transnational corporations, which, of course, control and manipulate Congress, the administration, and the Pentagon. For more about these intimate relationships, I refer you to the "Iron Triangle" chapter in *Missile Envy*.

The standard of journalism these days is quite low. *USA Today* epitomizes the "sound bite" mentality. Like TV sound bites, its articles are short, superficial, and lacking in substance, and they have lots of colored pictures, like comic books. Frequently, I am told by TV commentators that they do not have time to properly investigate a story, because of time lines, but we depend on the media to educate and guide us so that we may save the earth. Usually, I am given three minutes on the "Today Show" to tell the American people about the medical effects of nuclear war, while some film star is given eight. Important news is trivialized, and the rich, famous, or beautiful people are worshiped. This sort of reporting insults the American public.

For a nation of its size there are relatively few newspapers in the United States which offer sophisitcated journalism. And even with these one must learn to read between the lines and to think critically. I heard a former CIA agent, openly admit in the International Court of Justice that the CIA places "disinformation" stories (overt lies about foreign affairs that then influence U.S. political decisions) on the front page of the *New York Times,* the *Washington Post,* and others, as do the right-wing think tanks. Disinformation segments are also produced for TV.

The U.S. media now reach into almost every country on earth, through satellite TV, video tapes, printed magazines, and newspapers. But the culture of Hollywood is not appropriate for the people of Fiji, New Guinea, or Africa. These populations see ads for Pepsi or Coke, with their subliminal or overt messages of affluence, complete with cars and the "good life," and they want it. They listen to the music and the lyrics, and their culture is degraded by comparison with the glitz. The whole world is becoming "deculturalized" into a uniform "Coca-Cola society," wanting and needing an American way of life.[24] This is a terribly dangerous development because if 5.2 billion people lived the way the inhabitants of Hollywood do, the earth would be destroyed within the next fifty to a hundred years. Remember that the typical U.S. citizen pollutes 20 to 100 times more than the average Third World person. Furthermore, the earth does not have the resources to sustain 5 billion people in affluence. The rich must decrease their standard of living so that the other 4 billion may survive and prosper.

The Australian Broadcasting Corporation (ABC) is an autonomous body funded by the Australian government—answerable to no one, with no corporate sponsorship, similar in structure to the British Broadcasting Corporation (BBC). It is staffed by excellent investigative journalists, and one of the main reasons I came home from the United States was to be able to tune into the independent ABC.

Obviously, if the world is to survive, the press must not be used as a profit-making venture for a few people who are rich beyond compare and who now almost control the world. The place to start breaking down the corporate structure is the United States. If Americans all used their democracy appropriately, they would force members of Congress, who legally represent them, to legislate against private media ownership. Until they take the law into their own hands, they will be controlled by transnational corporations, which tell them only what they want them to know and which will amuse and numb them out with trivia and superficial "entertainment," violence, sex, and sport. I think all the major media—TV, newspapers, and radio—should not be privately owned but operated along the lines of the ABC and the BBC.

Walter Cronkite was moved to give a spontaneous outpouring in a speech soon after the election of George Bush. He said that the democratic loss in the election "was the fault of too many who found their voices stilled by not-so-subtle ideological intimidation." He continued,

> For instance, we know that unilateral military action in Granada and Tripoli was wrong. We know that Star Wars means uncontrollable escalation of the arms race. We know that the real threat to democracy is in half the nation in poverty.
>
> We know that Thomas Jefferson was right when he said 'A democracy cannot be both ignorant and free! We know that no one should tell a woman she has to bear an unwanted child. We know that religious beliefs cannot define patriotism.
>
> We know that it is not how one's lips are formed, but what they say. And we know there is freedom to disagree with all or part of what I've just said.
>
> But God Almighty, we've got to shout these truths in which we believe from the rooftops.[25]

10

Healing the Planet: Love, Learn, Live, and Legislate

The only cure is love. I have just walked around my garden. It is a sunny, fall day, and white fleecy clouds are scudding across a clear, blue sky. The air is fresh and clear with no taint of chemical smells, and the mountains in the distance are ringed by shining silver clouds. I have just picked a pan full of ripe cherry guavas to make jam, and the house is filling with the delicate aroma of simmering guavas. Figs are ripening on the trees and developing that gorgeous deep red glow at the apex of the fruit. Huge orange-colored lemons hang from the citrus trees, and lettuces, beetroots, and cabbages are growing in the vegetable garden. The fruit and vegetables are organically grown, and it feels wonderful to eat food that is free of man-made chemicals and poisons.

It is clear to me that unless we connect directly with the earth, we will not have the faintest clue why we should save it. We need to have dirt under our fingernails and to experience that deep, aching sense of physical tiredness after a day's labor in the garden to really understand nature. To feel the pulse of life, we need to spend days hiking in forests surrounded by millions of invisible insects and thousands of birds and the wonder of evolution. Of course, I realize that I am very fortunate indeed to be able to experience the fullness of nature so directly—literally in my own backyard. For many people—especially those living in urban areas who are unable to travel out of them regularly—such an experience is difficult to come by. Still, I urge all to try in some way to make a direct connection with the natural world.

Only if we understand the beauty of nature will we love it, and only if we become alerted to learn about the planet's disease processes can we decide to live our lives with a proper sense of ecological responsibility. And finally, only if we love nature, learn about its ills, and live accordingly will we be inspired to participate in needed legislative activities to save the earth. So my prescription for action to save the planet is, Love, learn, live, and legislate.

We must, then, with dedication and commitment, study the harm we humans have imposed upon our beloved earth. But this is not enough. The etiology of the disease processes that beset the earth is a by-product of the collective human psyche and of the dynamics of society, communities, governments, and corporations that result from the innate human condition.

We have become addicted to our way of life and to our way of thinking. We must drive our cars, use our clothes dryers, smoke our cigarettes, drink our alcohol, earn a profit, look good, behave in a socially acceptable fashion, and never speak out of turn or speak the truth, for fear of rejection.

The problem with addicted people, communities, corporations, or countries is that they tend to lie, cheat, or steal to get their "fix." Corporations are addicted to profit and governments to power, and as Henry Kissinger once said, "Power is the ultimate aphrodisiac."

The only way to break addictive behavior is to love and cherish something more than your addiction. When a mother and a father look into the eyes of their newborn baby, do they need a glass of beer or a cigarette to make them feel better? When you smell a rose or a gardenia, do you think of work or do you forget for a brief, blissful moment everything but the perfection of the flower? When you see the dogwood flowers hovering like butterflies among the fresh green leaves of spring, do you forget your worries?

Now, try to imagine your life without healthy babies, perfect roses, and dogwoods in spring. It will seem meaningless. We take the perfection of nature for granted, but if we woke up one morning and found all the trees dying, the grass brown, and the temperature 120°F, and if we couldn't venture outside because the sun would cause severe skin burns, we would recognize what we once had but didn't treasure enough to save.

To use a medical analogy: we don't really treasure our good health until we lose it or experience a dreadful accident. When I am injured, I always try immediately after the trauma, psychologically to recapture the moment before, when I was intact and healthy. But it is too late.

It is not too late, though, for our planet. We have ten years of work to do, and we must start now. If we do not, it may be too late for the survival of most species, including, possibly, our own. "But what can I do?" I hear you ask. Let me tell you.

In the industrialized world, and indeed in most of the Third World now, governments are more and more run and organized by a few corporations. But the corporatization of government is not conducive to global survival, as we have seen. A corporate mentality encourages greed, selfishness, and consumerism, not compassion for people or for nature. How, then, do we, the so-called little people or the grass roots, destroy this corporate stronghold of our country's politicians and governments so that we can develop a kind, compassionate society?

Well, I have some suggestions based on the Australian experience. The Australian system has many flaws that need correction, but in several important ways it provides a good model for America.

COMPULSORY VOTING

In Australia, voting is compulsory—if people do not vote, they are fined $50. I always get a feeling of pride as I stand at the polling booth watching my compatriots, old and young, rich and poor, fat and thin, black and white, healthy and infirm, roll or stagger up to the booth to vote. They do it with a sense of responsibility and dignity that befits a healthy democracy. Mandatory voting invokes responsibility—knowing that they have to vote, people have thought carefully about what they are voting for.

The result from a recent state election was rather interesting. The opinion polls had the premier of New South Wales winning by a large majority, but in fact the wishes of the people did not follow the polling data. The premier was just barely reelected, and he can obtain a majority to govern only by working with several independent candidates who hold the balance of power. The premier of New South Wales, a graduate of Harvard Business School, had run the government as a profit-making enterprise, by cutting rail services to rural areas and by cutting funding for the public school system, thereby enraging the rural sector as well as teachers and parents. The government went into the election confident, but the people of New South Wales were incensed because the premier had displayed no compassion for the little people, and consequently he almost lost his mandate to govern. No doubt, during his last term of office, the rich got richer, but then they are not the majority, and with a compulsory voting system, they could not prevail.

In the United States, in many elections, under 50 percent of the eligible citizens vote. Indeed, Reagan was swept into office in 1980 in a landslide victory, although only 27% of the voting-age population voted for him. I believe that most Americans feel that they can have little, if any, influence upon their rather corrupt politicians and system of government; anyway, they think, "What difference will *my* vote make?"

COMPULSORY REGISTRATION

In Australia, registration is also compulsory, and almost everyone is a registered voter. In the United States, registration has been made difficult for the average person. Rules and laws pertaining to voter registration differ from city to city and state to state. Poor people, especially, find it hard to travel to the place of registration in many locations because of inadequate public transport, let alone to understand the numerous forms and to comply with the regulations and conditions.

The historical reason for this grossly unfair system is clear: to limit the franchise. In order to remove the power and privileges of the wealthy and corporate elite, America needs compulsory registration and compulsory voting so that the political will and needs of all the people will be represented.

There is a simple and cheap way to achieve uniform voter registration in the United States. Since the address of every citizen is known by the U.S. Postal Service, registration could be effected by governmental decree.

CAMPAIGN FUNDING

I strongly suggest that political campaigns be completely funded by the federal government. When I ran for the Australian Parliament, in March 1990, I received ninety-one cents from the government for every vote I received. I spent a total of $40,000, raised $13,000 from individual donations from my electorate, and contributed $13,000 from my own funds, and the government paid $13,000.

It is vital to pass legislation that prohibits funding of political campaigns by special-interest groups and corporations, because by the time most politicians get elected in the States, they have prostituted themselves to the organizations that financed their extremely expensive campaigns. It now costs over $4 million to get elected to the House of Representatives—compared with $40,000 for a similar Australian campaign—$6 million to the Senate, and $60 million or more to the presidency. This tends to exclude from the Congress any person who is not a millionaire or who is not a gifted fund-raiser. In this situation, ordinary people can never participate in the governance of their country.

OLD-STYLE CAMPAIGNING

Politics is not really politics any more. It is run, for the most part, by Madison Avenue advertising firms, who sell politicians to the public the way they sell bars of soap or cans of beer. The campaigns are monitored by polls, and the whole event takes on an artificial, contrived quality. Aspiring politicians promote their images by appearing for thirty-second "sound bite" TV slots, surrounded by their attractive, socially acceptable family, smiling a big smile while mouthing inane phrases. Or the ads viciously attack the opponent, often by means of lies, slander, and other dirty tactics. These TV ads are outrageously expensive, and that helps account for the high cost of political campaigns.

My political campaign lasted three weeks. It began when I had lunch with the editor of the local newspaper, who tried to persuade me to run by saying that Parliament needed more women. I was not at all responsive to his line of reasoning, until he began to discuss the Gorbachev peace initiatives. Then he asked me, "What is Australia doing to support Gorbachev?" Then the light turned on. I realized that the last twenty years of my life had been devoted to educating people about the medical consequences of nuclear war. I saw that this led logically and inexorably to running for Parliament in order to provide more effective support for the global peace movement.

I drove home from the luncheon with a sense of excitement and anticipation. I campaigned in my rural electoral district of seventy thousand by holding big public meetings in all the towns and cities. People were very responsive, and they raised many intelligent and thoughtful questions. We discussed and debated various issues, and we all learned from the experience.

As it happened, I lost the election by only 684 votes, but in the process I defeated the incumbent, who was the leader of the National party—the most conservative political party in Parliament. The political pundits received a terrible shock, and Australia has not been the same since.

I therefore urge a return to old-style campaigning in the United States, one where politicians actually enter into a dialogue with their electorate, where political issues are debated, and where subjects affecting the fate of the earth are seriously discussed. Eliminate TV and radio advertising, and stop politicians from condescending to and staying aloof from the public, so that ordinary people cease feeling ostracized and can have input into their government.

Never forget that your politicians are not your leaders, not even the president. You are their leaders, and they are your representatives.

PROPORTIONAL REPRESENTATION

The next step in the formation of a truly representative government is proportional representation. Women make up 51.3 percent of the U.S. population, yet they have relatively little power or authority in the political system. The future of the world is in grave danger, and men continue to make the decisions, most of which have little relevance to saving the planet. I venture to suggest that if the composition of Congress were altered by a constitutional amendment mandating that half of the members of Congress be women, the world would be a very different place. Of course, the 50 percent would include good women and bad, stable and neurotic, smart and not so smart—as does the average mix of male politicians who now dominate Congress.

As we have seen, women also account for about half of the earth's population, yet they do two-thirds of the work. This toil and responsibility has gained them virtually no political power. Those few, like Margaret Thatcher and Indira Gandhi, who do rise to the top need to emulate the most powerful and ruthless male behavior. It is women who maintain their intrinsic feminine qualities whom the world desperately needs.

But women must also give credit to their own intellectual abilities and realize that they are at least as intelligent as most men—sometimes more so. We need to assume our place in the political sun and take over the job of helping to steer the planet toward a safe future for our children.

We must all develop the sense that we are an integral part of our governmental structure. There are many ways to achieve this sense. For instance, a young mother with three children and a part-time job will have little time to be deeply involved in political campaigns. But then again, she needs to guarantee her children's future. So she can join an organization that I founded in 1980, called Women's Action for Nuclear Disarmament (WAND), which is now the most effective lobbying body in Congress against nuclear weapons production. Our members are so well versed in the subject that even the most powerful hawks in Congress defer to them before they draft legislation concerning nuclear weapons and delivery systems. Before WAND and the allied Professionals Coalition for Nuclear Disarmament (composed of physicians, lawyers, educators, and scientists), only the military-industrial organizations and the Pentagon helped set the legislative agenda.

WAND has chapters in many towns and cities, so you can either join an established chapter or start one yourself. The address for WAND is P.O. Box B, Arlington, MA 02174. WAND will help you find out who your federal and state representatives are, what committees they belong to, how they vote, and how you can influence their voting. As you come to understand the political process better, and as your children grow up and you have more spare time, you may decide to run for local, state, or federal posts yourself. If you are a support person and not necessarily a leader, you will develop the political campaign for other women and give them the sort of strategies and nurturing that they need. Let's set a goal of 50 percent representation by the year 2000.

Of course, there are millions of good men who care deeply about the earth. If you are one of them, you are very welcome to join WAND. It needs sensitive, compassionate men who will transform political parties by their loving, caring behavior and who are courageous enough to speak the truth with a passion rarely heard in the corridors of political power or, for that matter, in those of corporate power.

POLITICIANS AND THEIR ELECTORATES

In 1989, I visited Iceland as a guest speaker at the fiftieth-anniversary commemoration of women's voting rights. Iceland is a fascinating, quite isolated country. Geologically, it is a volcanic island composed almost solely of black basalt rock, which supports very little life except moss, lichens, and tundra. Hot springs bubble from beneath the rocks, volcanoes hiss and spurt lava, and geysers soar majestically into the sky. Two massive tectonic plates meet in the middle of Iceland, and they are always moving ever so slowly. In such a place of planetary commotion, one gets the sense of a living, moving, evolving, magical earth.

The politics of Iceland are less turbulent but just as fascinating. In the center of Reykjavik, the capital city, the prime minister and politicians congregate at lunchtime in the public saunas that are heated by the underground hot springs. These eminent people sit in the baths alongside people from the general public, old and young, and conduct daily political conversations and dialogues. All issues are discussed, and the politicians return to work after lunch, having been briefed by their public.

Had I won office in March 1990 and entered Parliament, I would have taken my political directives from my electorate. I would have flown home every weekend from Canberra, the capital city, and conducted a public meeting in a different town or city on each occasion. I would have reported to my constituents about the preceding and following week's legislative activities, given them my opinion, and asked for their input. After suitable discussion and debate, I would have returned to Parliament armed with a consensus to guide and direct my activities and voting. I believe this is the only fair and legitimate way democracy should be conducted.

I ran as an independent candidate; I received no money from special-interest groups and was therefore answerable only to my electorate. The two major political parties in Australia are basically corrupt because they receive large corporate campaign contributions and because the system of party loyalty allows no dissent or independence on the part of individual representatives. All members become a cog in the party machine, which is run by several tough fellows who metaphorically kick in heads if people do not conform.

In the United States, too, it is abundantly clear that most, though not all, politicians do not have the luxury or freedom to truly represent the wishes of their constituents, because they have already sold out almost before they are elected. Furthermore, both the Democrats and the Republicans are, for the most part, fundamentally conservative. As Gore Vidal once said, "America has one political party with two right wings."

It has by now become obvious that the politics of the past and present are no longer appropriate models for the politics of the future, if the earth is to survive. I therefore propose that independent green candidates run in every congressional district and for every Senate seat in future elections. Some will almost surely win, and even if just three or four independents enter the House of Representatives, the potential for change in the political system of the United States will be very real. The independents by their very presence will cause a sea change in the thinking and behavior of the Democratic and Republican parties. At the very least, the urgency of the task to rescue the planet will be a part of the official agenda. No longer will CFC legislation compromise with the fate of the earth or with global warming, no longer will nuclear reactors be allowed to operate, producing hideous toxins to poison future generations, and no longer will corporate

America be allowed to interfere with government agendas.

Independent greens should also run for local and state government, a signal that the citizens of the United States have decided to become healers of the planet in their own right.

Elections will be fought and won in the public arena, organized with big public meetings and political debates, door knocking, outdoor rallies, and energetic performances by the candidates. Huge political campaign funds will be rendered unnecessary once the candidates meet their constituents face to face instead of on television. The politics of America will then become truly democratic again in the style of Lincoln and Jefferson.

I hesitate at this stage to suggest the formation of a national green party, because organizations of any sort tend to begin with revolutionary, fresh thinking but, with growing success, tend to become mired in bureaucracy and power struggles. This is old-style politics. Rather, I see the candidate and the electorate as forming a separate, autonomous unit that takes its seat in the Congress or the Senate and that votes as its own bloc, free of external influences. With such an arrangement, we will really have representational democracy.

This is not a pipe dream. The vote for independent green candidates is on the increase in Australia and indeed worldwide, and the United States needs to catch up with this new ecological and political trend. There is great hope and excitement in this development, and I encourage you to become *the* candidate of the future in *your* congressional district. It is time for the second American Revolution. If Soviet citizens can rebel, so can Americans.

We all have powers and talents that are unique and that the earth needs. We no longer have the right to hide our light under a bushel; the world needs our God-given talents, whatever they may be, and if we all decide to become autonomous and powerful in our own right, we will save the earth.

Remember though, we must start now, this instant, for there is no time to waste. We have only the next ten years, and each moment is precious. Always stay in the light, always be hopeful, and if obstacles arise, step over them, through them, or under them. If you have a sense of destiny and some knowledge that good will prevail over evil, we will save the earth.

Hope for the earth lies not with leaders but in your own heart and soul. If you decide to save the earth, it will be saved. Each person can be as powerful as the most powerful person who ever lived—and that is you, if you love this planet.

Notes

INTRODUCTION

1. "What the U.S. Should Do," *Time*, Jan. 2, 1989, 65.
2. Tara Bradley-Steck, "Expert: Americans Cause Much Damage to Planet," *Philadelphia Inquirer*, April 6, 1990.
3. World Commission on Environment and Development, *Our Common Future* (New York: Oxford University Press, 1987), 102.

1. OZONE DEPLETION AND THE GREENHOUSE EFFECT

1. Julian Cribb, "CFC Limits Will Take 14 Years to Reverse UV Trend," *Australian*, Sept. 10, 1991.
2. United Nations Environment Program (UNEP), *Action on Ozone* (1989).
3. Michael D. Lemonick, "Deadly Danger in a Spray Can," *Time*, Jan. 2, 1989, 42.
4. UNEP, *Action on Ozone*.
5. Ibid.
6. Larry B. Stammer, "Ozone Aid for Third World Slow to Arrive," *Los Angeles Times*, April 15, 1991.
7. "A Greenpeace Australian Strategy to Protect the Ozone Layer" (Greenpeace, P.O. Box 51, Balmain 2041, Australia, 1990).
8. Monica Oppen, *There's a Hole in My Ozone, Dear Liza* (Sydney: Atmospheric Project, Friends of the Earth, 1989).

9. Cribb, "CFC Limits."
10. Thomas E. Graedel and Paul I. Crutzen, "The Changing Atmosphere," *Scientific American,* Sept. 1989, 58–68.
11. "Greenpeace Australian Strategy."
12. "U.S. Launch Vehicles Claimed Harmful to Ozone Layer," *Moscow Aviatsiya Rosmonavtika,* Dec. 6, 1989.
13. NPR's "Weekend Edition," May 25, 1991.
14. William J. Broad, "NASA Moves to End Longtime Reliance on Big Spacecraft," *New York Times,* Sept. 16, 1991.
15. Sharon Ebner, "Solid Fuel Critics Say Ozone Is in Danger," *Sun Herald* (Mississippi), March 17, 1990.
16. UNEP, *Action on Ozone.*
17. "U.S. Launch Vehicles"; Walter H. Corson, ed., *The Global Ecology Handbook: What You Can Do About the Environmental Crisis* (Boston: Beacon Press, 1990), 230.
18. UNEP, *Action on Ozone.*
19. Margaret Harris, "Skin Cancer out of Control," *Sydney Morning Herald,* Sept. 12, 1990.
20. UNEP, *Action on Ozone.*
21. Ibid.
22. Ibid.
23. Helen Thew, "Thinning Threat to Ozone Layer," *Daily Telegraph,* Sept. 22, 1990.
24. "Greenpeace Australian Strategy."
25. Stammer, "Ozone Aid for Third World."
26. UNEP, *Action on Ozone.*
27. Ibid.; Lemonick, "Deadly Danger in a Spray Can."
28. United Nations Environment Program, *The Greenhouse Gases* (Nairobi, 1987).
29. Lester R. Brown et al., *State of the World, 1990: A Worldwatch Institute Report on Progress toward a Sustainable Society* (New York: W. W. Norton, 1990), 18.
30. Stephen Schneider, "The Changing Climate," *Scientific American,* Sept. 1989, 70–79.
31. Mike Seccombe, "An Ill Wind That Only Does Cows Good," *Sydney Morning Herald,* July 18, 1989.
32. Brown et al., *State of the World, 1990,* 17.
33. "Another Culprit of the Greenhouse Effect: Jet Aircraft," *Sydney Morning Herald,* Aug. 27, 1991.
34. UNEP, *Greenhouse Gases;* Schneider, "Changing Climate"; Michael Lemonick, "Feeling the Heat," *Time,* Jan. 2, 1989.

35. Phillip Shabecoff, "Cloudy Days in Study of Warming World Climate," *International Herald Tribune*, Jan. 19, 1989.
36. Schneider, "Changing Climate."
37. Corson, ed., *Global Ecology Handbook*, 232.
38. Lemonick, "Feeling the Heat."
39. Ibid.
40. Ibid.
41. Corson, ed., *Global Ecology Handbook*, 233.
42. Ibid., 232.
43. UNEP, *Greenhouse Gases*.
44. Schneider, "Changing Climate."
45. UNEP, *Greenhouse Gases*.
46. Corson, ed., *Global Ecology Handbook*, 233.
47. UNEP, *Greenhouse Gases;* Schneider, "Changing Climate."
48. Schneider, "Changing Climate."
49. UNEP, *Greenhouse Gases*.

2. ATMOSPHERIC DEGRADATION: CAUSES AND SOME SOLUTIONS

1. United Nations Environment Program (UNEP), *Action on Ozone*.
2. Greenpeace Media Release, Balmain, NSW, Australia, June 20, 1990.
3. Jonathan Kwitny, "The Great Transportation Conspiracy," *Harper's*, Feb. 1981, 14–21.
4. "What the U.S. Should Do," *Time*, Jan. 2, 1989, 65.
5. Jeremy Leggett, ed., *Global Warming: The Greenpeace Report* (New York: Oxford University Press, 1990), .
6. Walter H. Corson, ed., *The Global Energy Handbook: What You Can Do About the Environmental Crisis* (Boston: Beacon Press, 1990), 192.
7. Lester R. Brown et al., *State of the World, 1990: A Worldwatch Institute Report on Progress toward a Sustainable Society* (New York: W. W. Norton, 1990), 120.
8. Corson, ed., *Global Ecology Handbook*, 192; Leggett, ed., *Global Warming*, 261, 262.
9. Leggett, ed., *Global Warming*, 269.
10. Ibid., 289.
11. Corson, ed., *Global Ecology Handbook*, 205.
12. *Ibid*
13. Helen Caldicott, *Missle Envy* (New York: Morrow, 1984), 208.
14. Brown et al., *State of the World, 1990*, 26.

15. Ibid., 128.
16. Corson, ed., *Global Ecology Handbook,* 203–4.
17. Ibid.
18. Ibid.
19. *Missile Envy*, 208.
20. *Infact Brings GE to Light* (Boston: Infact, 1988), 38–39.
21. Jeff Gerth, "Contractors' Role at Energy Dept. Called Pervasive," *New York Times,* Nov. 6, 1989.
22. *Nuclear Power and National Energy Strategy* (Washington, D.C.: Nuclear Information and Resource Service, April 1991).
23. Richard Lacayo, "Global Warming: A New Warning," *Time,* April 22, 1991, .
24. Ibid.
25. Brown et al., *State of the World, 1990,* 34.
26. Leggett, ed., *Global Warming,* 231.
27. Ibid., 25.
28. Ibid., 24, 25.
29. World Commission on Environment and Development (WCED), *Our Common Future* (New York: Oxford University Press, 1987), 193.
30. Ibid., 192.
31. Corson, ed., *Global Ecology Handbook,* 93.
32. Ibid., 205; WCED, *Our Common Future,* 200.
33. Sting and Jean-Pierre Dutilleux, *Jungle Stories: The Fight for the Amazon* (London: Barrie & Jenkins, 1989), 4.
34. Brown et al., *State of the World, 1990,* 25.
35. WCED, *Our Common Future,* 199.
36. Lester R. Brown et al., *State of the World, 1990,* 36.
37. Ibid., 37.
38. Ibid., 35, 36.

3. TREES: THE LUNGS OF THE EARTH

1. Edward O. Wilson: "Threats to Biodiversity," *Scientific American,* Sept. 1989, 108–16.
2. Michael Kennedy, "Endangered," *Habitat* (Australian Conservation Foundation, Fitzroy, Victoria), Aug. 1990, 4–7.
3. Walter H. Corson, ed., *The Global Ecology Handbook: What You Can Do About the Environmental Crisis* (Boston: Beacon Press, 1990), 235.

4. Ibid., 124.
5. Sting and Jean-Pierre Dutilleux, *Jungle Stories: The Fight for the Amazon* (London: Barrie & Jenkins, 1989), 9, 31.
6. William K. Stevens, "Research in Virgin Amazon Uncovers Complex Farming," *New York Times,* April 3, 1990.
7. Ibid.; Sting and Dutilleux, *Jungle Stories,* 78.
8. Sting and Dutilleux, *Jungle Stories,* 71.
9. "How Your Tax Dollars Fill the Rainforests" (advertisement by Rainforest Action Network, San Francisco), in *New York Times,* Oct. 15, 1990.
10. Ibid.
11. Sting and Dutilleux, *Jungle Stories,* 38, 40, 51.
12. Corson, ed., *Global Ecology Handbook,* 124.
13. Sting and Dutilleux, *Jungle Stories,* 15.
14. Corson, ed., *Global Ecology Handbook,* 120.
15. Petra Kelly, "The Environmental Crises" (Address delivered at the Ninth IPPNW World Congress, Hiroshima, Oct. 9, 1990).
16. Associated Press, "Rainforest Being Wiped Out to Make Chopsticks," *Sydney Morning Herald,* April 15, 1989.
17. Sting and Dutilleux, *Jungle Stories,* 39, 41.
18. Ibid., 41.
19. Ibid., 45.
20. Ibid., 39.
21. Physicians for Social Responsibility, *Our Common Future, Healing the Planet: A Resource Guide to Individual Action* (Los Angeles, 1989).
22. Sting and Dutilleux, *Jungle Stories,* 43.
23. Douglas Farah, "Cocaine Lords Confound the Amazon Catastrophe," *Australian,* Jan. 12–13, 1991.
24. Sting and Dutilleux, *Jungle Stories,* 10.
25. Corson, ed., *Global Ecology Handbook,* 118.
26. Eugene Linden, "The Death of Birth," *Time,* Jan. 2, 1989, 32–35.
27. Sting and Dutilleux, *Jungle Stories,* 20.
28. Bruce Rensberger, "Scientists See Signs of Mass Extinction," *Washington Post,* Sept. 29, 1986. The Club of Earth includes Paul Ehrlich, Edward O. Wilson, Ernst Mayer, and Thomas Eisner.
29. Sting and Dutilleux, *Jungle Stories,* 33.
30. Ibid.; Wilson, "Threats to Biodiversity"; Corson, ed., *Global Ecology Handbook,* 117, 118.
31. Corson, ed., *Global Ecology Handbook,* 122.
32. Ibid., 117.

33. Michael Richardson, "South East Asian Nations Move to Protect Forests," *International Herald Tribune,* Jan. 12, 1989.
34. Corson, ed., *Global Ecology Handbook,* 124.
35. Ibid., 225, 228; World Commission on Environment and Development, *Our Common Future* (New York: Oxford University Press, 1987), 180.
36. Corson, ed., *Global Ecology Handbook,* 228.
37. Ibid., 225.
38. *Conservation News* (Australian Conservation Foundation), 22, no. 5 (June 1990).
39. E. Schwartz, "A Proportionate Mortality Ratio Analysis of Pulp and Paper Mill Workers in New Hampshire," *British Journal of Industrial Medicine* 45 (1988): 234–38.
40. Keith Schneider, "U.S. Backs Off Dioxin's Deadly Ranking as Ecologists Protest," *International Herald Tribune,* Aug. 16, 1991.

4: THE WITCHES' CALDRON: TOXIC POLLUTION

1. Walter H. Corson, ed., *The Global Ecology Handbook: What You Can Do About the Environmental Crisis* (Boston: Beacon Press, 1990), 272.
2. Ibid.
3. Alan B. Durning, "The Quick-Food Addiction," *Arizona Daily Star,* Sept. 17, 1991.
4. Cheri Sanders, "City Life: USC Environmental Medicine Researchers Measure the Risks of Nuclear Living," *USC Medicine,* Winter, 1991.
5. Eric Mann, "Lighting a Spark: L.A.'s Smogbusters," *Nation,* Sept. 17, 1990, 257, 268–74.
6. United Nations Environment Program, *Hazardous Chemicals,* UNEP Environment Brief no. 4.
7. United Nations Environment Program, *Industry and the Environment,* UNEP Environment Brief no. 7.
8. UNEP, *Hazardous Chemicals.*
9. Ibid.; UNEP, *Industry and the Environment.*
10. Tara Bradley-Steck, "Expert: Americans Cause Much Damage to Planet," *Philadelphia Inquirer,* April 6, 1990.
11. Corson, ed., *Global Ecology Handbook,* 247.
12. Ibid., 251.

13. John Langone, "A Stinking Mess," *Time,* Jan. 2, 1989, 44–47.
14. Corson, ed., *Global Ecology Handbook,* 248.
15. William Glaberson, "Love Canal: Suit Centers on Records from 1940's," *New York Times,* Oct. 22, 1990.
16. Ibid.
17. Corson, ed., *Global Ecology Handbook,* 248.
18. "Taking on the Nation's Biggest Polluters" (20/20 Vision, California Sixteenth District, P.O. Box 5781, Monterey County, CA, Sept. 1991).
19. Lenny Seigel, Garry Cohen, and Ben Goldman, "The U.S. Military's Toxic Legacy" (National Toxic Campaign Fund, 1168 Commonwealth Ave., Boston, MA, 1991).
20. Ibid., 249.
21. Ibid., 251.
22. Ibid.
23. Ibid.
24. Ibid.
25. Marlise Simons, "Nowhere to Hide?" *St. Louis Post Dispatch,* April 11, 1990.
26. Ibid.
27. Ibid.
28. Timothy Egan, "A Lonely Law Enforcer Pursues New Violator," *New York Times,* 1990.
29. Corson, ed., *Global Ecology Handbook,* 252.
30. Ibid.
31. Ibid.
32. "Race on to Go Organic," *Northern Star* (Lismore, Australia), Dec. 19, 1990.
33. Corson, ed., *Global Ecology Handbook,* 253.
34. Ibid., 253–54.
35. UNEP, *Hazardous Chemicals.*
36. Corson, ed., *Global Ecology Handbook,* 250.
37. Timothy Egan, "Goo Galore," *New York Times Book Review,* April 28, 1991.
38. Keith Schneider, "Judge Rejects $100 million Fine for Exxon in Oil Spill as Too Low," *New York Times,* April 25, 1991.
39. Thomas C. Hayes, "Earnings Soar 75% at Exxon," *New York Times,* April 25, 1991.
40. David Beers and Catherine Capellaro, "Greenwash," *Mother Jones,* March–April 1991, 38, 39, 40, 41, 88.

41. Paul Brown, "Critical Look in the Mirror for OPEC," *Sydney Morning Herald,* Jan. 29, 1991.
42. Eric Karlstrom, "Nobody's Hanging Yellow Ribbons for the Persian Gulf," *Sierra Runoff,* April 1991.
43. "Kuwait, Drilling New Wells, Says Iraq Damage Nears End," *New York Times,* Sept. 15, 1991.
44. John Miller, "Scientists Preview Environmental Effects of a Gulf War" (International Clearing House on the Military and the Environment/ARC, P.O. Box 150753, Brooklyn, NY, Jan. 14, 1991); Michael Parrish, "The Spoils of War," *Los Angeles Times Magazine,* June 23, 1991; Matthew L. Wald, "No Global Threat Seen from Oil Fires," *New York Times,* June 25, 1991.
45. Matthew L. Wald, "No Global Threat Seen from Oil Fires," *New York Times,* June 25, 1991.
46. Ibid.
47. "Deadly Black Tide May Be Unstoppable," *Sydney Morning Herald,* Jan. 30, 1991.
48. Deborah Smith, "Oil Threatens Dugongs, Survivors of Another War," *Sydney Morning Herald,* Jan. 30, 1991.
49. Parrish, "Spoils of War."
50. Ibid.
51. Sean Ryan, "Oil Spill Poses Threat to Turtles," *Australian,* Feb. 18, 1991.
52. Parrish, "Spoils of War."
53. Miller, "Scientists Preview Environmental Effects of Gulf War."
54. Ibid.
55. John Horgan, "Science and the Citizen: Up in Flames," *Scientific American,* May 1991, 24.
56. Ibid., 17.
57. Karlstrom, "Nobody's Hanging Yellow Ribbons."
58. Ibid.
59. Stanley Meisler, "Pollution and Deep Blue Sea," *Los Angeles Times,* Oct. 20, 1990.
60. Mark Tran, "Poor Training Blamed for Soviet Sub Mishaps," *International Herald Tribune.*
61. William M. Arkin and Joshua M. Handler, "Nuclear Disasters at Sea, Then & Now," *Bulletin of the Atomic Scientists,* July–Aug. 1989.
62. Corson, ed., *Global Ecology Handbook,* 200.
63. Susan Wyndham, "Death in the Air," *Australian Magazine,* Sept. 29–30, 1990; Keith Schneider, "Opening the Record on Nuclear Risks," *New York Times,* Dec. 3, 1989; Kenneth B. Noble, "The U.S. for

Decades Let Uranium Leak at Weapon Plant," *New York Times,* Oct. 15, 1988.

64. Wyndham "Death in the Air."
65. Ibid.
66. Matthew L. Wald, "Chemicals in Hanford, A Plant Create Risk of Explosion, Study Says," *New York Times,* Oct. 17, 1989.
67. Helen Caldicott, *Nuclear Madness: What You Can Do* (New York: Bantam, 1980), 42.
68. Martha Odom, "Tanks That Leak, Tanks That Explode . . . Tanks Alot DOE," *Portland Free Press,* Vol. 1, #5.
69. Matthew L. Wald, "Wider Peril Seen in Nuclear Waste from Bomb Making," *New York Times,* March 28, 1991.
70. Wyndham "Death in the Air."
71. Larry Lang, "Missing Hanford Documents Probed by Energy Department," *Seattle Post-Intelligencer,* Sept. 20, 1991.
72. Wyndham, "Death in the Air."
73. Keith Schneider, "Seeking Victims of Radiation Near Weapon Plant," *New York Times,* Oct. 17, 1988.
74. Noble, "U.S. for Decades Let Uranium Leak at Weapon Plant."
75. Schneider, "Opening the Record on Nuclear Risks."
76. Keith Schneider, "Brain Cancer Cases in Los Alamos to Be Studied for Radiation Link," *New York Times,* July 23, 1991.
77. Matthew L. Wald, "In Nuclear Cleanup, Costs Grow and Promises Fade," *New York Times,* Oct. 29, 1989.
78. Brad Knickerbocker, "Huge Cleanup Awaits Arms Plants," *Christian Science Monitor,* March 15, 1991.
79. Matthew L. Wald, "The Adventures of the Toxic Avengers Have Barely Begun," *New York Times,* Sept. 15, 1991.
80. Corson, ed., *Global Ecology Handbook,* 196, 201.
81. Walter Ellis, Vera Rich, and James Blitz, "The Truth about Chernobyl," *Australian Magazine,* June 16–17, 1990.
82. "Chernobyl Update," KPFA (Los Angeles), April 30, 1991.

5. SPECIES EXTINCTION

1. Eugene Linden, "The Death of Birth," *Time,* Jan. 2, 1989, 32–35.
2. Edward O. Wilson, "Threats to Biodiversity," *Scientific American,* Sept. 1989, 108–16.
3. Ibid.
4. Ibid.

5. Ibid. 114.
6. World Commission on Environment and Development, *Our Common Future* (New York: Oxford University Press, 1987), 4.
7. Walter H. Corson, ed., *The Global Ecology Handbook: What You Can Do About the Environmental Crisis* (Boston: Beacon Press, 1990), 102.
8. Ibid.; Robert Lamb, "The Displeased Coral Diver Was Small Beer," *International Herald Tribune*, Jan. 19, 1989; Mike Seccombe, "Government Divided over Reef Oil Exploration Plans," *Sydney Morning Herald*, July 18, 1990; Greg Roberts, "Resort Sewerage Killing Coral Says Expert," *Sydney Morning Herald*, March 7, 1989; Deborah Smith, "30 Years Isn't Long to Save the Species," *Sydney Morning Herald*, Aug. 19, 1990.
9. Michael Kennedy, "Endangered," *Habitat*, Aug. 29, 1990.
10. Roger Beckman and Steven Morten, "Where Have All the Desert Mammals Gone?" *Wildlife* (Wildlife Preservation Society of Queensland, Brisbane), Spring 1990
11. Corson, ed., *Global Ecology Handbook*, 103.
12. Ibid.
13. "Help WWF Stop the Rhino Horn Trade" (World Wildlife Fund Campaign Report, April 1991).
14. CITES The Convention on International Trade in Endangered Species of Wild Fauna and Flora, United Nations Environment Program, UNEP Environment Brief no. 8, P.O. Box 30552, Nairobi, Kenya; Smith, "30 Years Isn't Long to Save the Species"; Charles P. Wallace, "A Wildlife Smugglers Paradise," *Los Angeles Times*, Oct. 20, 1990.
15. Corson, ed., *Global Ecology Handbook*, 103.
16. Richard Cole, "Amphibians in Global Decline Say Scientists," *Northern Star*, June 20, 1990; Stan Ingram, "Of Fire, Water, Earth and Air—The Mystery of the Disappearing Frog," *Wildlife*, Spring 1990.
17. David Charley, "Famous Finches Feel the Pressure," *Northern Star*, June 30, 1990.
18. Jane Brody, "Studies Point to Food Web Danger," *Sydney Morning Herald*, Feb. 15, 1989; "A Strategy for Antarctic Conservation: International Union for Conservation of Nature and Natural Resources, IUCN—The World Conservation Union" (Proceedings of the Eighteenth Session of the IUCN General Assembly, Perth, Australia, Nov. 28–Dec. 5, 1990).
19. Ibid.
20. Peter Pringle, "The Green Bottles Don't Accidently Fall on a US Heap," *Sydney Morning Herald*, July 28, 1989.

21. Paul Grigson, "Wall of Death Headed for Southern Waters," *Sydney Morning Herald,* July 8, 1989.
22. Ibid.
23. Paul Grigson and Judith Whelan, "Japan Bows to Pressure against Drift Net Fishing in the South Pacific," *Sydney Morning Herald,* July 18, 1990.
24. Judith Whelan, "Drift Net Dolphin Toll Put at 6,400," *Sydney Morning Herald,* June 23, 1990.
25. Brian Woodley, "Moratorium Fails to Stop Killing of Whales," *Weekend Australian,* June 30–July 1, 1990.
26. Dianne Dumanowski "Measles-like Virus Reported in Dolphins," *Boston Globe,* October 24, 1990.
27. Ibid.
28. Timothy Egan, "Environmentalists Flipping over Dolphins Use as Navy Guard Dogs," *Sydney Morning Herald,* April 15, 1989.
29. Personal communication from the Naval Oceans Systems Center, San Diego.
30. Associated Press, "Greenpeace Wants Ban on Fiji Turtle Shells," *Northern Star,* July 5, 1990.

6. OVERPOPULATION

1. Toni Carabillo and Judith Meuli, "The Feminization of Power" (Fund for the Feminist Majority, Los Angeles, 1988).
2. Walter H. Corson, ed., *The Global Ecology Handbook: What You Can Do About the Environmental Crisis* (Boston: Beacon Press, 1990), 54.
3. World Commission on Development and Environment, *Our Common Future* (New York: Oxford University Press, 1987), 109.
4. Robert Steinbrook, "Gains Reported in Test of Male Contraceptive," *Los Angeles Times,* Oct. 20, 1990.
5. Louise Silvestre et al., "Voluntary Interruption of Pregnancy with Mifepristone (RU 486) and a Prostaglandin Analogue," *New England Journal of Medicine* 322 (March 8, 1990): 645–48.
6. Corson, ed., *Global Ecology Handbook,* 32.
7. Ibid., 33.
8. Ibid., 52.
9. Private 1991 communication with Cuban consul general, Sydney, Australia.
10. F. E. Trainer, *Abandon Affluence!* (London: Zed Books, 1985), 176.

11. Corson, ed., *Global Ecology Handbook,* 30–31.
12. Ibid., 52.
13. Ibid., 53.

7. FIRST WORLD GREED AND THIRD WORLD DEBT

1. F. E. Trainer, *Abandon Affluence!* (London: Zed Books, 1985), 116.
2. Walter H. Corson, ed., *The Global Ecology Handbook: What You Can Do About the Environmental Crisis* (Boston: Beacon Press, 1990), 44.
3. Ibid., 71.
4. Trainer, *Abandon Affluence!* 75, 117.
5. Corson, ed., *Global Ecology Handbook,* 70.
6. Ibid., 72.
7. "Doctors Link Red Meat Diet and Colon Cancer," *Northern Star,* Jan. 11, 1991.
8. Corson, ed., *Global Ecology Handbook,* 73.
9. Ibid., 28, 73, 77.
10. Ibid., 73.
11. Trainer, *Abandon Affluence!* 118, 142.
12. Corson, ed., *Global Ecology Handbook,* 74–76, 81.
13. Ibid., 58.
14. Trainer, *Abandon Affluence!* 157.
15. Ibid., 158.
16. Ibid., 146, 157; Corson, ed., *Global Ecology Handbook,* 59.
17. Corson, ed., *Global Ecology Handbook,* 57.
18. Trainer, *Abandon Affluence!* 177.
19. Corson, ed., *Global Ecology Handbook,* 58.
20. Ruth Leger Sivard, *World Military and Social Expenditures* (Leesburg, Va: World Priorities, Washington 1987).
21. Alameda County SANE/FREEZE, "Taxes" (1987).
22. Ibid.
23. Susan George, "A Fate Worse Than Debt" (BBC documentary, 1990).
24. Ibid.
25. Ibid.
26. Ibid.
27. Ibid.
28. Ibid.

29. Corson, ed., *Global Ecology Handbook,* 45.
30. Ibid., 45, 46; George, "Fate Worse Than Debt."
31. George, "Fate Worse Than Debt."
32. Carol Sherman, "A Look Inside the World Bank" (Rainforest Information Centre, Byron Bay, NSW, Australia, 1989).
33. Corson, ed., *Global Ecology Handbook,* 45, 46.
34. George, "Fate Worse Than Debt."
35. Trainer, *Abandon Affluence!* 141.
36. Corson, ed., *Global Ecology Handbook,* 46.
37. George, "Fate Worse Than Debt."
38. Trainer, *Abandon Affluence!* 152–53.
39. Ibid., 153.
40. George, "Fate Worse Than Debt."
41. Trainer, *Abandon Affluence!* 143.
42. George, "Fate Worse Than Debt."
43. Corson, ed., *Global Ecology Handbook,* 46–47.
44. George, "Fate Worse Than Debt."
45. Ibid.
46. Corson, ed., *Global Ecology Handbook,* 47.
47. Ibid., 44.
48. Ibid., 46.
49. "10 Million Face Starvation Just in the Horn of Africa," *Sydney Morning Herald,* Dec. 21, 1990; Reuters, "Massive Starvation Looms in Africa," *Northern Star,* Dec. 19, 1990.
50. Reuters, "1990 Was the Warmest Year Yet," *Northern Star,* Jan. 11, 1991.
51. "10 Million Face Starvation."
52. Reuters, "Massive Starvation Looms in Africa."
53. Corson, ed., *Global Ecology Handbook,* 68.
54. Ibid., 68, 77.
55. Trainer, *Abandon Affluence!* 153.
56. World Commission on Environment and Development, *Our Common Future* (New York: Oxford University Press, 1987), 123.
57. Trainer, *Abandon Affluence!* 165.
58. Greg Clough and Ted Wheelwright, *Australia: A Client State* (Ringwood, Australia: Pelican Books, 1982), 12.
59. Trainer, *Abandon Affluence!* 233.
60. Ibid., 173.
61. "GATT, the Environment and the Third World" (Michelle Syverson & Associates, Resource Guide Environmental News Network, 1442a Walnut St., Berkeley, CA 94709, 1990).

62. Chakravarthi Raglaven, *Recolonization* (Penang, Malaysia: Third World Network, 1990), 37, 63.
63. Ibid., 91.
64. Ibid., 45.
65. Clough and Wheelwright, *Australia,* 4–15.
66. Raglaven, *Recolonization,* 95.
67. Corson, ed., *Global Ecology Handbook,* 80.
68. "GATT, the Environment and the Third World."
69. Raglaven, *Recolonization,* 98.
70. Trainer, *Abandon Affluence!* 143–44.
71. "GATT, the Environment and the Third World."
72. Associated Press, "Ban on Tuna Imports Held to Violate Treaty," *Washington Post,* Aug. 26, 1991.
73. Ibid.
74. Kenichi Ohmae, "Toward a Global Regionalism," *Wall Street Journal,* April 27, 1990.
75. Associated Press, "Big Macs a Hit with Soviets," *Sydney Morning Herald,* Feb. 2, 1991.

8. THE MANUFACTURE OF CONSENT

1. Alex Carey, "Managing Public Opinion" 1983–1987; idem, "From Cominfirm to Capinform: A Bipartisan Approach to 1984" 1983–1987; idem, "Conspiracy or Groundswell" 1983–1987; idem, "Business Propaganda and Democracy" 1983–1987. Copies of these unpublished works by Alex Carey can be obtained from Mr. Lou Kiefer, Western Regional Officer, International Machinists and Aerospace Workers, P.O. Box 1400, Oakland, CA 94604.
2. A. L. Lowell, *Public Opinion in Popular Government* (New York: Longman Green, 1926), 43.
3. George Brown Tindall, *America: A Narrative History,* 2d ed. (New York: W. W. Norton, 1988), 990.
4. *The Penguin English Dictionary* (Harmondsworth, England: Penguin Books, 1985–86).
5. Tindall, *America,* 1005.
6. [Bernays].
7. Tindall, *America,* 1006.
8. Lasswell, Propaganda Techniques in World War One, M.I.T. Press, Cambridge, 1971, 222.

9. "National Association of Manufacturers" Proceedings of the 40th Annual Convention of the Congress of American Industry, 1935, 25.
10. S. H. Walker and Paul Sklar, *Business Finds Its Voice: Management's Effort to Sell the Business Idea to the Public* (New York: Harper, 1938), 202.
11. K. Sward, "The Johnstown Steel Strike of 1937," in *Industrial Conflict,* ed. G. W. Hartman and T. Newcombe (New York: Corden, 1939), 74–102.
12. Richard S. Tedlow, "The National Association of Manufacturers and Public Relations during the New Deal," *Business History Review* 50 (1976), 25–45.
13. Rippa, "Organized Business and Public Education."
14. Daniel Bell, "Industrial Conflict and Public Opinion," in *Industrial Conflict,* ed. A. W. Kornhauser, R. Dubin, and A. M. Ross (New York: McGraw-Hill, 1954), 240–56.
15. Morris Bartel Schnapper, ed., *The Truman Program: Addresses and Messages* (Washington, D.C.: Public Affairs Press, 1968), 84–85.
16. Peter F. Drucker, "Have Employer Relations Had the Desired Effect?" (American Management Association Personnel Services, no. 134, 1959).
17. Walter H. Corson, Ed., *The Global Ecology Handbook: What You Can Do About the Environmental Crisis* (Boston: Beacon Press, 1990), 44.
18. F. H. Knelman, *America, God & the Bomb* (Vancouver, New Star Books, 1987).
19. A. Crittendon, "A New Corporate Activism in the US," *Australian Financial Review,* July 18, 1978.
20. Kim McQuaid, "The Round Table Getting Results in Washington," *Harvard Business Review* 59 (May–June 1981): 114–23.
21. Green and Buchshaum, *The Corporate Lobbies: Political Profiles of the Business Roundtable and the Chamber of Commerce* (Public Citizen, Washington, 1980), 15.
22. John S. Saloma III, *Ominous Politics—The New Conservative Labyrinth* (New York: Hill and Wang, 1984), 14–15.
23. Leonard Silk, "Some Things Are More Vital Than Money When It Comes to Creating the World Anew," *New York Times,* Sept. 22, 1991.
24. Martin Mayer, "A 'Family Man' Who Bled the Weak, Meek," *Sydney Morning Herald,* Jan. 21, 1991, excerpted from Martin Mayer, *The Collapse of the Savings and Loan Industry* (New York: Scribner's, 1991).
25. Ibid.

26. John Durie, "US Banking Reforms Skirt Sensitive Issues," *Australian*, Feb. 7, 1991.
27. "US Treasury Urged to Examine Insurance," *Weekend Australian*, Dec. 29–30, 1990.
28. "Junk Bond Takeover Boom Ends," *Weekend Australian*, Dec. 29–30, 1990.
29. Robert Pear, "Balanced Budget: Soon a $362 Billion Mirage," *International Herald Tribune*, Aug. 17–18, 1991.
30. Kate Legge, "George Bush: A Man Obsessed," *Weekend Australian*, Feb. 2–3, 1991.
31. Pear, "Balanced Budget."
32. Anthony Ramirez, "Coca-Cola Earnings Set Record," *New York Times*, April 18, 1991.
33. Jason De Parle, "Crux of Tax Debate: Who Pays More?" *New York Times*, Oct. 15, 1990.
34. "A Balanced Use of Our Taxes," *Community Concept Care* (Auburn, Maine), 4, no. 5 (Oct.–Nov. 1990).
35. Deborah Cameron, "Your Money or Your Life," *Sydney Morning Herald*, July 8, 1989.
36. Ibid.
37. Henry A. Waxman, "Frenzy of Cuts in Medicine Plugs Deficit," *Los Angeles Times*, Oct. 9, 1990.
38. Cameron, "Your Money or Your Life."
39. *Geography! An International Gallup Survey* (Princeton, N.J.: Gallup Organization, 1988).

9. AMERICAN MEDIA AND THE FATE OF THE EARTH

1. Reuters, "Six Mega-media Giants by Year 2000: Study," *Northern Star*, Feb. 2, 1991; Ben Bagdikian, "The 26 Corporations That Own Our Media," *Extra* (New York), June 1987.
2. Reuters, "Six Mega-medial Giants."
3. *INFACT Brings GE to Light* (Boston: INFACT, 1988).
4. "The Bill Casey Connection" (editorial), *Extra*, March–April 1990.
5. Doug Henwood, "Capital Cities/ABC: No. 2, and Trying Harder," *Extra*, March–April 1990.
6. Doug Henwood, "The Washington Post: The Establishment's Paper," *Extra*, Jan.–Feb. 1990.

7. Ibid.
8. Doug Henwood, "Corporate Profile: The New York Times," *Extra,* March–April 1989.
9. Peter Dykstra, "Polluters' PBS Penance," *Extra,* May–June 1990.
10. William Hoynes and David Croteau, "All the Usual Suspects, MacNeil/Lehrer and Nightline," *Extra,* Winter 1990.
11. "Study Aftermath: Fair Debates MacNeil/Lehrer," *Extra,* May–June 1990.
12. Ibid.
13. Ibid.
14. "Businessmen: TV's Oppressed Minority," *Extra,* June 1987; Doug Henwood, "Public's TV's Elite Market," *Extra,* Summer 1990.
15. Jonathan Tasini, "Lost in the Margins: Labor and the Media," *Extra,* Summer 1990.
16. Ibid.
17. Ibid.
18. Doug Henwood, "Media Activists Target Minnesota Public Radio Corporate Profile," *Extra,* March–April 1989.
19. David S. Broder, "The Insiders' Outrage Falls Short of Journalism's Mark," *International Herald Tribune,* Jan. 12, 1989.
20. Helen Caldicott, *Missile Envy* (New York: Bantam Books, 1986), 271–77.
21. "ABC's 1984 Coverup for the Gipper," *Extra,* March–April 1990.
22. Anthony Lewis, "Mr. Murdoch's Shadow," *New York Times,* Nov. 5, 1987.
23. Ibid.
24. Greg Clough and Ted Wheelwright, *Australia: A Client State* (Ringwood, Australia: Pelican Books, 1982), 33.
25. Jeremy Gerard, "Walter Cronkite: This Is the Way It Is," *International Herald Tribune,* Jan. 10, 1989.

Index

INDEX

INDEX

Index